IMPACT EXTRUSION OF METALS

Impact and Cold EXTRUSION *of* METALS

by

JOHN L. EVERHART, P.E.

Metallurgical Engineer

CHEMICAL PUBLISHING COMPANY, INC.
NEW YORK
1964

Printed in the United States of America

Preface

The impact or cold extrusion of metals is a " chipless machining " process that can greatly reduce costs of material and labor if properly applied. Within certain limitations in shape, some parts can be produced by this process that often require only trimming to length. On others, the secondary operations are reduced greatly over those required for several of the competing processes. Finally, certain components can be made that can be produced by no other method.

Impact extrusion can be employed for the production of parts from a wide range of metallic materials. In commercial production at present are components produced from aluminum, copper, lead, magnesium, tin, zinc and their alloys and various steels. In the development stage are parts produced from less common metals, such as titanium and molybdenum.

The aim of this book is to present a coordinated description of impact extrusion. Its purpose is to present the information required by an engineer to determine whether the process is applicable to his products, and to assist the metallurgist in selecting the metal to be used.

The book is based on personal experience, talks, and correspondence with producers of components, and a search of the free-world technical literature. It is divided into two parts. The first deals with the underlying principles and the characteristics of the process; the second, with the metals being extruded and the products.

It is a pleasure to express my thanks to all those that assisted in developing the theme. I feel particularly indebted to T. J. Dolan, who prepared a number of the sketches; to my son, Larry, who assisted in the preparation of the manuscript; and to my wife, Helen, who prepared the graphs and many of the sketches, and assisted in proofreading.

John L. Everhart

Westfield, New Jersey
January, 1964

Contents

Chapter 1

INTRODUCTION

No term has been evolved that is universally recognized, as defining the process to be discussed in this book. Although the original application in the production of collapsible tubes—is over a century old, the process is still called by various names. These include *impact extrusion*, *cold extrusion*, *extrusion forging*, *cold pressing*, *extrusion pressing*, and *cold forging*. Of these terms, *impact extrusion* is used most frequently by producers of nonferrous parts; and *cold extrusion* is preferred by producers of ferrous parts.

It has been said that to define a process is to know it. About thirty years ago, the German Stamped Metal Institute defined this process as,

> *Kaltspritzen*, using a punch and die, is the process of transforming a blank in the cold state into a tubular part, during which the material is forced to flow under a high, rapidly applied pressure.[1]

The literal translation of *kaltspritzen* is "cold-squirting." In 1942, the name was changed from *kaltspritzen* to *kaltfliesspressen* (literally "cold flow-pressing") without changing the definition. Although the latter term, or simply *fliesspressen*, is generally used today in German publications, *kaltspritzen* is still used to some extent.

In the United States and Great Britain, *impact extrusion* was the original name for the process. It implied that the part was formed by forcing metal to flow through an orifice by means of an impact blow. This term is not strictly accurate. Soft metals, at least, can be forced through an orifice merely by squeezing, provided a sufficiently powerful press is used. The term *impact extrusion*, however, has long been established as a specific method of production, readily distinguished from hydraulic extrusion and forging.

The term *cold extrusion* was introduced with the application of impact extrusion to ferrous metallurgy. Although the product is an extrusion through an orifice, the advocates of cold extrusion state that the optimum speed is not great enough to justify calling it *impact extrusion*. Libergeld,[2] who adapted the process to steel, did not agree with this conclusion. In the original German patent, he stated that

> the invented process, herewith described, eliminates the undesirable features of existing cold- and hot-forming of hard metals by the application of the *kaltspritzen* process used in the manufacture of hollow, nonferrous shapes to similar shapes made of steel.

There is some justification in selecting the term, *cold extrusion*, to identify the process. The operation is cold working, if such working be defined as that conducted below the recrystallization temperature of the metal. It should be noted that, even though the extrusion is started with blanks at ambient temperature, resistance to flow is sufficient to raise the temperature of the part and of the tools several hundred degrees. The rise in temperature with most metals, however, is not sufficient to reach the recrystallization temperature, and parts are produced in a work-hardened condition. Exceptions are lead and tin which undergo hot working at relatively low temperatures.

Sachs[3] has called this process *extrusion-forging*. Because the walls are extruded and the bottom of the shell is forged, this term is probably the most descriptive that has been proposed. It eliminates "impact" and "cold," and offers the possibility of a further distinction based on the kind of working involved. Thus, the process could be subdivided into cold extrusion-forging and hot extrusion-forging in a manner analogous with cold rolling and hot rolling. The term, *extrusion-forging* however, has not received much support as a name for the process.

Regardless of the various names used to describe it, basically the same process is involved, and progress in the application of this technique would be stimulated if a single name could be chosen to characterize it. As this has not occurred, the terms most widely accepted, *impact extrusion* and *cold extrusion*, will be used in this book, interchangeably. In general, however, *impact extrusion* will be used in discussing nonferrous parts; and *cold extrusion*, in discussing ferrous parts.

GENERAL NATURE OF THE PROCESS

The methods used in impact extrusion are shown, schematically, in Figures 1:1 and 1:2. The two processes were called, respectively, *impact extrusion* and *Hooker extrusion* for many years. Current practice in most plants seems to be to call them *backward-* and *forward extrusion*, and this practice will be followed in this book.

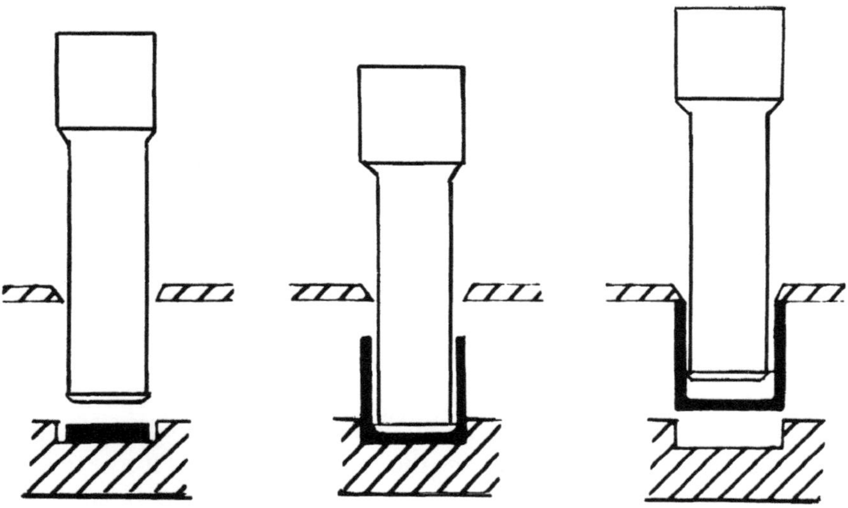

FIG. 1:1. BACKWARD EXTRUSION
Left: Blank in die chamber; *center*: extrusion of shell; *right*: stripping shell

Backward Extrusion

The original method of impact extrusion is not known as *backward extrusion*. In this method, a blank of the material to be extruded is placed in the cavity of a die and struck by a punch. The metal of the blank is forced to flow upward through the annular opening between punch and die and it rises along the punch. The thickness of the wall of the part is determined by the clearance between punch and die. The thickness of the base is determined by the stroke of the press and is independent of the thickness of the wall. The operation is shown schematically in Figure 1:1.

Backward extrusion can be used to produce a wide variety of shells with circular, oval, square, rectangular, and various other cross sections. In general, the side walls must be perpendicular to the base to permit flow during extrusion, but ribs can be formed on internal or external walls. In addition, multi-wall shells can be formed.

Backward extrusion has the advantages of simplicity of tooling and ease in removing the shell after forming. There are also limitations. As the punch must be long enough to accommodate the entire length of the part to be extruded, plus an allowance for trimming, a long-stroke press is frequently required. There is a limit to the length that can be extruded by this process. In the extruding of long pieces, the punch has a tendency to float—a tendency that is increased by play in the moving head of the press. As a result, there is some difficulty in maintaining uniform thickness of the walls.

Forward Extrusion

The second method of impact extrusion is generally called *forward extrusion* although the original term—Hooker extrusion—is still used to some extent. In this method, a blank placed in the cavity of a die, having a constriction as indicated in Figure 1:2, is struck by a punch that fits so closely that metal cannot escape between the punch and the walls of the die. As the punch descends, it compresses the blank and causes it to extrude through the orifice formed by the projection on the die and the nose of the punch. Forward extrusion resembles the direct extrusion of hot billets in hydraulic presses but the length of the part that can be formed by forward extrusion is more restricted.

Although flat blanks can be used, blanks are generally preformed to the contour necessary to obtain the desired configuration in the extruded part. Forward extrusion can be used to produce open- or closed-end tubes; the bottoms of closed-end tubes can be the same thickness as the side-walls or heavier. Side walls can be parallel or tapered. Parts can be produced with fillets.

Forward extrusion requires less power than backward extrusion, and smaller presses can be used to produce parts of equivalent size. As true alignment of punch and die is assured, tolerances can be held more closely than is possible in backward extrusion. Because the length of the part is practically independent of the length of the punch, a short stocky punch can be used.

There are disadvantages, also. Secondary tooling may be required to permit removal of the extruded part from the die and this requirement will increase the cost of tooling for forward extrusion over that required for a simple backward extrusion. In addition, the shapes that can be produced by forward extrusion are more limited than those that can be formed by backward

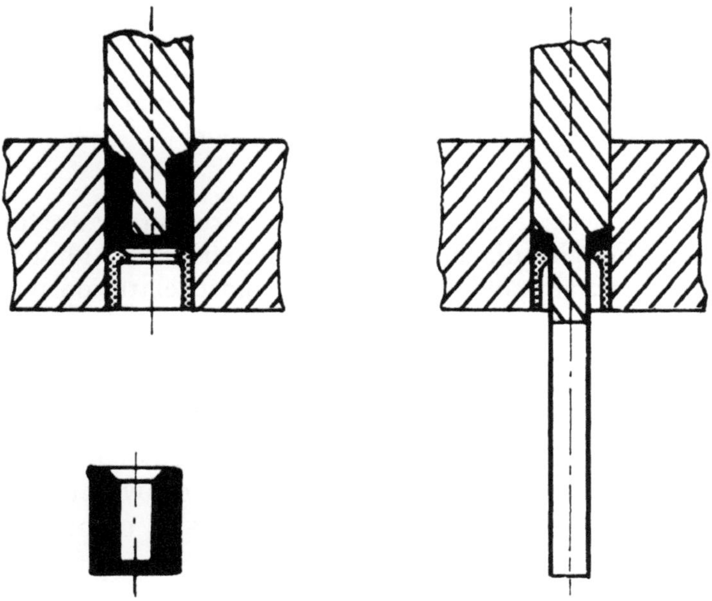

FIG. 1:2. FORWARD EXTRUSION
Lower left: Blank; *upper left*: blank in die chamber; *right*: extrusion of shell

extrusion. In particular, the base of the forward-extruded part must be plain because there is no forging action against the bottom of the die.

Combined Extrusion

From the preceding discussion, it may appear that each of the methods must be used separately. Such is not the case. Backward and forward extrusion can be used simultaneously to form parts and this procedure is followed frequently. Possibilities of combining the methods will be discussed in the chapter dealing with design.

COMPARISON OF IMPACT EXTRUSION WITH OTHER PROCESSES

The role of impact extrusion as a process in production can be indicated by comparison with certain other processes which may be considered to compete with it. These include deep drawing, hydraulic extrusion, and forging.

Deep Drawing

Both impact extrusion and deep drawing use a combination of punch and die to form a hollow shell from a cold blank. They differ in the manner in which deformation occurs. In impact extrusion, the metal is forced through an orifice under compressive stresses. In deep drawing, the metal is pulled into the die by the punch and stressed in tension.

Hydraulic Extrusion

Both impact extrusion and hydraulic extrusion involve forcing a metal through a die under compressive stresses. Although the metal is generally cold-worked during impact extrusion and hot-worked during hydraulic extrusion, this is not a distinguishing factor.

The major difference between the two processes lies in the type of product. That produced by impact extrusion is a single component which may or may not require additional operations to produce the finished part. The product of hydraulic extrusion is generally a rod, tube, structural shape, or other semifinished mill form.

Forging

There is no well-defined line of demarkation between backward extrusion and forging. The bottom of a backward extruded part is forged; in some press-formed parts, extensions, such as bosses, are extruded around the punch. Forgings are generally formed hot but can be formed cold in some metals, whereas impact extruded parts are usually cold-formed. Thus forging and impact extrusion overlap.

They differ in one characteristic. Forgings usually have confined recesses into which any extruded metal is forced. In impact extrusion, the extruded portion is not confined after passing through the orifice between punch and die.

Impact extrusion yields a more concentric product than

forging does and the parts produced are more uniform in weight. The surface of the part is also superior, thus reducing the number of finishing operations required.

ADVANTAGES AND LIMITATIONS OF IMPACT EXTRUSION

The economic significance of impact extrusion lies in the opportunities it affords for more effective use of materials, reduction of labor costs, elimination of intermediate operations, improvement in quality of product, and the attainment of high rates of production with relatively simple tools. There are also limitations. Some of the advantages and disadvantages are summarized in Table 1:1.[4] Comparison with other production processes points up the advantages more specifically.

TABLE 1:1

LIMITATIONS OF IMPACT EXTRUSION

Advantages	*Disadvantages*
1. High-speed production with low unit cost	1. Relatively few alloys can be processed commercially at present
2. Low scrap losses	2. Shapes that can be produced are limited
3. Tooling relatively cheap	3. Ratio of length to diameter is restricted
4. Bottom thickness independent of wall thickness	4. Maximum size of part is determined by capacity of press
5. Combines operations by forming ribs, fins, holes, during extrusion	5. Generally large production runs are required for economy
6. Parts can be made with zero degree draft angles	
7. Surface finishes of 20 to 70 micro-inches obtainable	
8. High mechanical properties can be obtained	
9. One-piece parts can be used instead of assemblies	

Deep Drawing

Selection of impact extrusion or deep drawing as the method of production often depends on the configuration of the part. In the production of some parts, the methods are competitive. Choice of impact extrusion may be based on the following advantages mentioned by Jevons[5] and Askew.[6]

1. Certain shapes such as deep thin-walled shells, which would be difficult to make by deep drawing, can be impact-extruded in one operation.

2. Certain metals such as lead, which cannot be deep drawn, can be easily impact-extruded into thin-walled shells.
3. The base of an impact-extruded part can be thicker than the side walls and shaped to various profiles—an operation that cannot be performed by ordinary deep drawing.
4. Tool costs to produce a deep shell are often lower for impact extrusion than for deep drawing because only a single set is required.
5. Higher rates of production can generally be maintained by impact extrusion.
6. It is often possible to use a thinner-walled impact-extruded part than a similar deep-drawn part because of the increased strength resulting from the severe cold working.
7. There is generally less scrap from impact extrusion than from deep drawing.

The differences in losses in scrap can be very significant. In German practice, the deep drawing of cups from aluminum, brass, or steel can result in 60 per cent of scrap; impact extrusion can reduce loss in scrap to 10 per cent.[7]

As a specific example of the advantages of impact extrusion over deep drawing, Gabler[1] cited the production of a hollow shell having a height six times the inside diameter. To produce this shell by deep drawing required blanking, three drawing stages, and leveling. The same part was produced in two steps by impact extrusion; cutting a blank from a rod, and backward extruding. The impact-extruded part required only 72 per cent as much material as the drawn part, and cost only 12 per cent as much.

Machining

The advantages of impact extrusion over machining have been discussed by Galloway.[8] Production of parts based on machining often results in losses of 50 to 60 per cent in scrap. With impact extrusion, these losses are reduced to about 10 per cent. Highly stressed parts that are produced by machining must be made from alloy steels that are subsequently heat-treated. Similar components can often be produced from low-carbon steels by impact extrusion because of the excellent mechanical properties developed by cold working.

Epner[9] cites the production of a control arm bushing for automotive use. This part, formerly produced on a screw machine and heat-treated to obtain the required properties, is now produced by cold extrusion and requires no heat treatment.

He mentioned also that impact-extruded parts have tolerances equal to those obtainable in conventional boring and turning operations, and that the excellent finish of the surface often avoids the need for grinding and polishing operations.

Other Processes

Impact-extruded parts have displaced castings, forgings, and weldments in various applications. Generally, the substitution has required redesign of a specific part. Examples will be given in the chapters dealing with the various metals that are being impact-extruded.

REFERENCES

1. Gabler, P., Precision Impact Extrusion and Die Casting of Aluminum in Industry, *Aluminium*, 25:194 (1943).
2. Liebergeld, A., German Patent 717679, issued February 20, 1942; effective August 27, 1935.
3. Sachs, G., *Sheet Metal Fabricating*, Reinhold Publishing Corp, New York, 1951.
4. Evfrhart, J. L., Impact (Cold) Extruded Parts, *Materials and Methods*, August 1955:111.
5. Jevons, J. D., *The Metallurgy of Deep Drawing and Pressing*, Chapman & Hall, London, 1940.
6. Askew, W. G., Progress in Impact Extrusion. The Production of Metal Tubes and Containers, *Metal Industry* (London), 54: 65 (1939).
7. Cold Extrusion and Deep Drawing, *Machinery* (London), 90:425 (1957).
8. Galloway, E. J., Some Experiments on the Impact Extrusion of Aluminum, Copper, and Steel, *Machinery* (London), 93:356 (1958).
9. Epner, E. J., Cold Extrusion, *Metal Progress*, October 1959: 122.

Part 1

THE PROCESS

Chapter 2

FUNDAMENTALS OF IMPACT EXTRUSION

Impact extrusion is a process in which metal is forced to flow under the influence of high compressive stresses. The mechanism depends on whether working occurs above or below the temperature of recrystallization. Wagner[1] has differentiated between these two types of activity by calling them the *thermoplastic* range and the *crystoplastic* range.

The thermoplastic range lies above the temperature of recrystallization and the structure of the crystals is not stable in this range. Thus, interatomic stresses, resulting from flow in this range, can be partially relieved by atomic rearrangement at the operating temperature. This rearrangement, however, requires a finite time interval. Lead and tin are in the thermoplastic range at room temperature, whereas aluminum, zinc, and magnesium are in the range at slightly elevated temperatures. Disturbance of the slip planes causes work-hardening, but is relieved to some extent by recrystallization—the actual relief depending on time and temperature.

The crystoplastic range is below the temperature of recrystallization. In this range, plastic flow causes permanent deformation of the structure with areas of relatively high energy at the slip planes. According to Wagner, there are three distinct regions in the crystoplastic range:

1. A low region near the yield point which has no sharply defined stress-strain relationship.
2. A middle region in which the true stress-strain relationship is sharply defined, i.e., the increase of stress with strain is fairly constant. This is the region in which major commercial forming operations are performed.

3. An upper region that is poorly defined. Transition into this region occurs with a sharp rise in stress with increasing strain. Templin[2] has defined the point of abrupt change in the stress-strain relationship as the point at which all favorably oriented slip planes have been exhausted. In the upper region, exceptional strength is developed during compression working operations, such as impact extrusion, and a rise in temperature resulting from internal friction appears to be essential.

Unlike the basic laws underlying elastic deformation, which are rather well defined, the laws of plastic deformation appear to be complex and are only partially understood. It is doubtful whether some of the physical laws are valid for material being worked in the plastic range. Wagner[1] points out that Hooke's Law seems to be inapplicable in the plastic range for a metal that hardens during working, whereas Pascal's appears to be only approximately correct. Poisson's ratio changes in value —a change that is to be expected because the more nearly a material approaches the liquid state, the nearer the ratio approaches a value of 0.5.

Because of the difficulties in developing the underlying principles, much dependence has been placed on experimental work. As happens frequently, however, practice has far outstripped theory, and much additional experimental work will be required before the laws of plastic deformation as applied to impact extrusion are fully understood. It is necessary to rely on data, many of which are qualitative, in discussing the factors that affect impact extrusion. Among these factors are flow during extrusion, extrusion forces, and speed of extrusion.

FLOW DURING IMPACT EXTRUSION

A number of investigations have been concerned with the determination of flow during impact extrusion, using either inspection of the finished part, marks scribed on the outside surfaces of the blanks, or split blanks, to obtain an indication of the process. The earliest investigations on lead and tin were run many years ago but, in general, the results were similar to those obtained in more recent work on other metals.

Boston[3] states, however, that aluminum does not have the same flow characteristics in backward extrusion as the softer metals have. According to him, lead and tin start to flow as soon as pressure is applied, whereas aluminum fills even the most minute scratch in the die as it is compressed initially. As a result, sufficient friction is developed to hold the skin momentarily. Extrusion, therefore, starts with metal flowing from the interior of the blank.

Whether or not this is a valid distinction will require additional investigation. It is certainly true that, in the backward extrusion of tin-coated lead blanks to form collapsible tubes, the tin and lead flow simultaneously and a tin-coated lead tube is produced.

It is also true that examination of a freshly extruded aluminum shell will show that the first metal extruded has come from the interior of the blank. If highly polished tools are used, the upper $\frac{1}{2}$ to $\frac{2}{3}$ of the shell has a mirror finish, whereas the lower $\frac{1}{3}$ to $\frac{1}{2}$ is dull, with the characteristic appearance of the original surface of the blank.

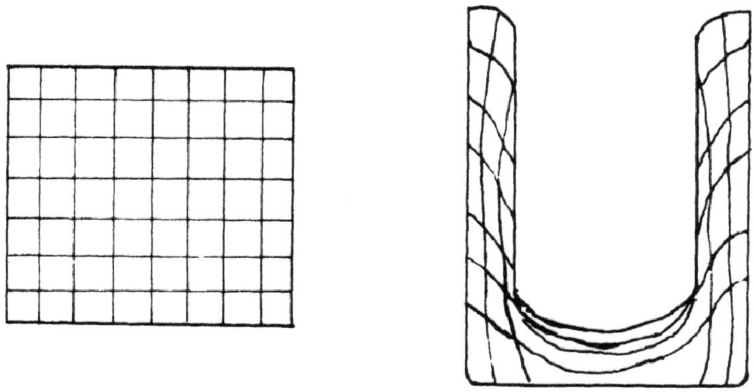

FIG. 2:1. BACKWARD EXTRUDED CARBON STEEL CUP SHOWING FLOW LINES RESULTING FROM EXTRUSION (*after* Feldmann[5])

Mabb[4] used a more precise method of determining the same flow properties. He scribed marks, perpendicular to the faces, on the edges of aluminum blanks, and different marks on the bottom surfaces. After backward extrusion, the marks from the edges were reproduced on the upper part of the shell, from the top down to about $\frac{1}{3}$ to $\frac{1}{2}$ the height. The marks, originally on the bottom surface of the blank, appeared near the bottom of the shell and extended only a short distance up from the bottom.

A number of investigators have used split blanks with grids scribed on the internal mating surfaces and have determined the mechanism of flow by studying the distortion of these grids after extrusion.

A typical example is that of Feldmann,[5] illustrated in Figure 2:1. This represents a 0.12%-carbon steel backward-extruded with a reduction of 40 per cent. The distortion occurring

FIG. 2:2. FORWARD EXTRUSION OF A SOLID BAR SHOWING FLOW LINES RESULTING FROM EXTRUSION (Hauttmann and Pessl[7])

during extrusion is apparent. The inhomogeneous nature of the deformation caused variations in hardness in both the longitudinal and the radial directions. Greatest work-hardening occurred near the internal surface and the bottom of the shell; least work-hardening occurred near the outside surface at the upper part of the shell.

Using a procedure similar to that of Feldmann, Sabroff and his associates[6] studied the flow of titanium blanks under both backward and forward extrusion. In backward extrusion, they found that the top surface of the blank flows and expands under the punch to form the inside surface of the extruded part and the layer of metal directly under the punch flows into the annular space between punch and die to form the initial portion of the wall. They decided that deformation at the outer surface occurs mainly by axial extension and, therefore, the inner surface is worked more severely than the outer surface.

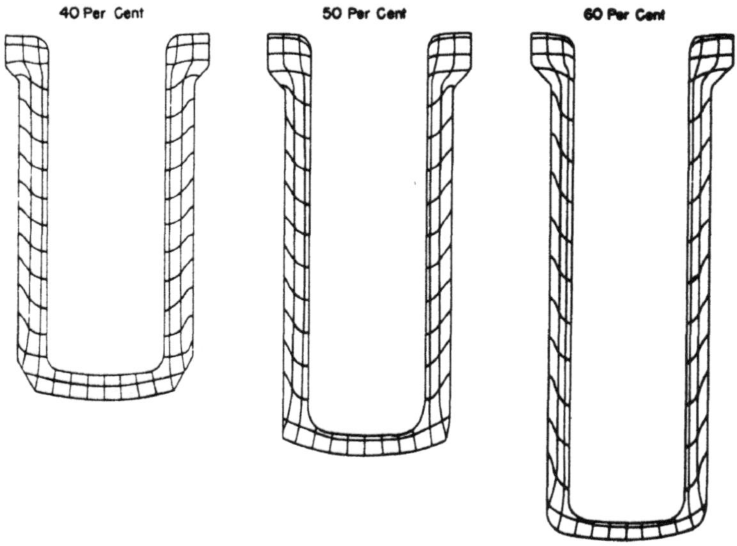

FIG. 2:3. FORWARD EXTRUSION OF HOLLOW PARTS SHOWING FLOW LINES RESULTING FROM EXTRUSION (Sabroff *et al*[6])

Hauttmann and Pessl[7] employed a split bar scribed with parallel lines to study the forward extrusion of solid parts. The split bar was reduced 70 per cent in the test. Flow lines showed that most of the deformation occurred in the core rather than at the outer surface of the part. A sketch of their results is shown in Figure 2:2.

In the forward extrusion of hollow parts, Sabroff *et al.*[6] found that the maximum deformation is in the radial direction. During the initial stages, the bottom of the blank moves into the die with little deformation. As extrusion proceeds, the flow becomes markedly less uniform because of shear stresses on the outer fibers. Deformation at the inner surface occurs chiefly by axial elongation and radial compression, with practically no shear even at high reductions of area. On the basis of these results they decided that, in contrast to backward extrusion, the outer surface is more severely worked in forward extrusion than is the inner surface.

Sabroff *et al.* studied split blanks after several different reductions. The shapes of the grid lines after three reductions in forward extrusion are indicated in Figure 2:3.

These investigators also measured hardness to check the validity of their deductions. They found that the inside surfaces of backward-extruded parts were from 5 to 25 Vickers numbers harder than the outside surfaces. The spread in values for hardness, however, decreased with increases in the reduction from blank to finished shell. In forward extrusion, by contrast, the outer surfaces were harder than the inner surfaces. The spread in hardness decreased from about 20 Vickers numbers for a reduction of 20 per cent to about 10 Vickers numbers for a reduction of 60 per cent.

The use of split blanks is open to the objection that flow may be influenced by the fact that two separate and distinct sections of metal are being treated as one. Therefore, a number of investigators have preferred to use a normal blank and section the shell for metallographic examination. The author prepared a series of backward extrusions of aluminum by varying the setting of the press, thus obtaining a group of specimens in which the development of the shell was carried in stages from merely compressing the blank to the completed shell. These specimens were sectioned, polished, etched, and examined metallographically. Although the angle of the die was 30 degrees, these specimens showed flow lines similar to those reported by Wagner for the direct hydraulic extrusion of parts through square orifices; i.e., there was a dead zone in the corner at the bottom of the blank. Movement of metal in the early stages occurred only in a region within a very short distance of the orifice; behind this region the granular structure was undisturbed. When considerable reduction was involved, however, as in making a thin-walled shell, the entire cross section of the wall showed flow lines.

Pineau[8] also used micrography to study flow during backward extrusion of a hollow part. He found that the longitudinal

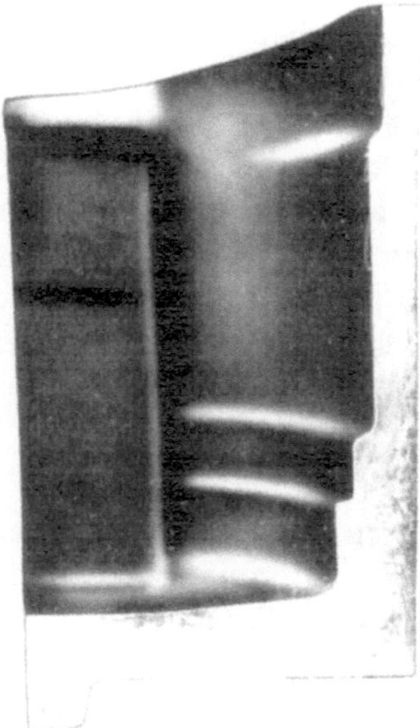

FIG. 2:4. ETCHED CROSS SECTION OF A SIMPLE BACKWARD EXTRUSION SHOWING FLOW LINES (Thompson Products, Inc.)

fibers in a simple shape curved toward the exterior wall. He observed that there was an unfibered section just above the base of the shell. This shell had a rather thick bottom and it appears that movement in this region depends on the degree of working.

FIG. 2:5. ETCHED CROSS SECTION OF A MORE COMPLICATED EXTRUSION SHOWING FLOW LINES (Magnesium Corporation of America)

The flow of material in backward extrusion is influenced by the shape of the part and it can be very complicated. The etched cross section of a rather simple backward-extruded part in Figure 2:4 shows the flow lines that developed in forming this part. Greatest flow has occurred near the inner wall whereas there are regions in the corners which show practically no movement. In an extruded part of more complex contour, as shown in Figure 2:5, the flow lines indicate that movement has occurred over most of the cross section.

EXTRUSION FORCES

Many equations have been developed for the relationship between deformation and requirements in force. One of the earliest was Sieber's[9] and it is pertinent to present some of his reasoning in developing this formula.

According to Sieber, the fundamental laws of cold plastic deformation are:

1. The volume of metal remains unchanged during change in shape.
2. The force and work requirements for cold forming depend on the degree of dimensional change in the three principal directions.

From these two laws, he derived a term which he called the *index of plastic deformation*. When applied to a cylindrical body, which is the most common shape in commercial practice, this index becomes

$$\varphi = \ln \frac{F_0}{F_1}$$

where φ = index of plastic deformation
F_0 = original cross-sectional area
F_1 = area after cold working

He showed, in addition, that the resistance to deformation during cold working depends on the composition of the material and its structure. This resistance to deformation is not a constant, and it varies with the degree of cold work unless the granular structure has been restored by annealing. Sieber termed the deformation resistance K_f; he showed that $K_f = f(\varphi)$, and that the relation could be expressed with considerable accuracy as

$$K_f = f(\varphi) = \sqrt[3]{3P \ln \frac{x_1}{x_0}} + \sigma_E$$

where σ_E = elastic limit
P = a constant for the material (obtained by testing)
x_0 = original length of the part
x_1 = final length of the part

These formulas apply to cold working, in general. Proceeding to plastic forming, Sieber states that K_f varies with the deformation, and that the work required depends on the integrated mean value of the resistance to forming, K_{fm}.

Turning now to the impact extrusion of hollow shells, Sieber showed that the total force required can be expressed by

$$P_{\max} = F_0 \cdot K_{fm}\left(\ln \frac{F_0}{F_1} + 0.6\right) \times \left(1.25 + \frac{\pi(D_0 + d)h\mu}{F_0}\right)$$

where F_0, F_1 = cross-sectional areas before and after extrusion (sq mm)
D_0 = diameter of die recess (mm)
d = diameter of punch head (mm)
h = internal height of hollow blank before forming (mm)
μ = coefficient of friction

Sieber also gave an approximate equation that is based on work by Dipper. This equation is

$$P_{\max} = F \cdot K_{fm}\left(2 + 0.25\frac{h_0}{s}\right) \cdot \frac{1}{\eta}$$

where F = cross-section of punch (sq mm)
h_0 = initial height of blank (mm)
s = wall thickness of extruded shell (mm)
η = degree of reduction (assumed to be between 0.5 and 0.8)

Many other formulas have been developed to predict pressures in extrusion, some of which appear to check more closely with actual measured pressures than the equations given above. As Chapter 3 deals with the factors affecting pressure in extrusion, the Sieber formulas will be compared with others in that chapter.

SHAPE OF THE EXTRUSION GRAPH

The oscillograph has been used frequently to investigate the shape of the extrusion graph and thus determine force required. There is general agreement that the extrusion force rises to a maximum as extrusion starts and then falls to some lower value. Kuhne[10] states that a second peak maximum occurs in thin-walled shells at the end of the stroke, but this does not appear in thick-walled shells. The second peak is greater for hard metals than for soft metals and it disappears completely in lead extrusions. It is usually lower than the maximum reached in the initial stages of extrusion.

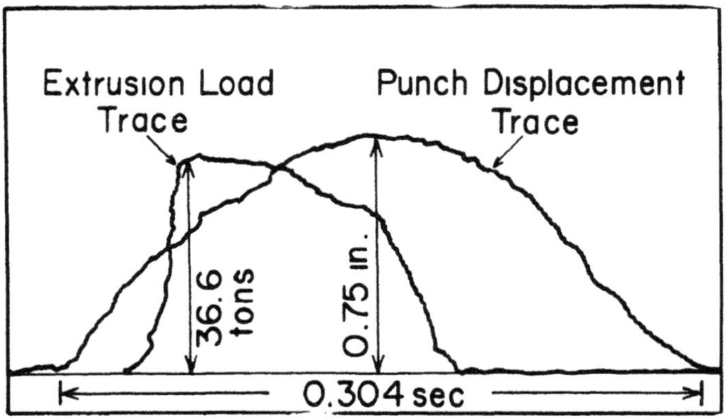

FIG. 2:6. TYPICAL TRACINGS OF PUNCH DISPLACEMENT AND LOAD TRACE DIAGRAM (*after* Galloway[11])

Figure 2:6 show typical tracings of punch displacement and extrusion load according to Galloway.[11] The load rises to the maximum in the very early stages of extrusion and decreases before the maximum displacement of the punch. In this graph, there is no indication of a second peak. Figure 2:7 shows how extrusion force varies with penetration of punch. This was obtained by superimposing diagrams made at different maximum penetrations of the punch, therefore showing different thicknesses at the base. The extrusion force rose rapidly to the maximum and then decreased slowly, until, depending on the ratio in the thickness of base to wall, it either fell sharply to zero or rose to a second peak before falling to zero. In this work, also, the second peak was less than the maximum force in the initial stages of the extrusion, as long as thicknesses of the base were maintained within practical limits.

Phillips[12] took motion pictures of the backward extrusion of an aluminum collapsible tube. The extrusion required 0.055 seconds measured from the moment the punch struck the blank to the complete formation of the tube. Study of the film showed an abnormal slowing of the rate of extrusion after about half of the extrusion had been completed.

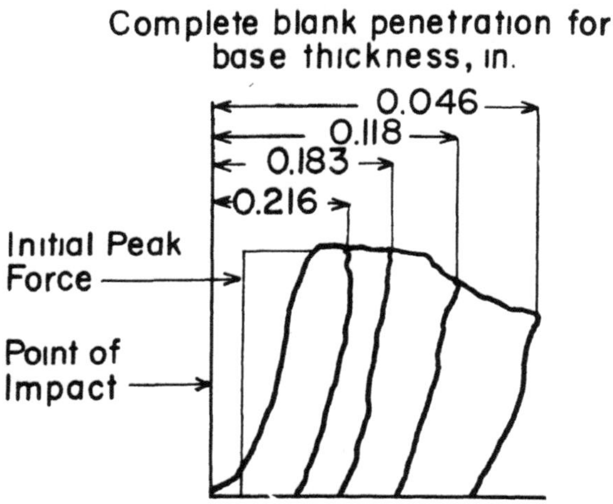

Fig. 2:7. Superimposed Oscillographs Showing Variation in Extrusion Force with Penetration (*after* Galloway[11])

He explained this phenomenon by using pressure curves determined with oscillographic recording equipment. The highest pressure during the extrusion occurred at the initial point of the cycle. This was followed by a rather rapid drop in loading until the extrusion cycle had been half completed.

Phillips believes that the rapid rate observed on the film during the first half of the extrusion cycle is the result of two mechanical actions, and that the rate during the last half of the cycle is the result of one action. At the high pressures reached during the start of extrusion, straining of the press causes elongation of the tie rods and frame. The combination of the harmonic motion imparted by the crank and the relief of the strains in the press frame causes high rates of extrusion at the beginning of the cycle. After the strains in the press are relieved, the rate of extrusion is governed by the harmonic motion of the extrusion press.

Boston[3] also states that there is a sharp rise in pressure to a maximum in the earliest stages of the extrusion. The pressure then tapers off gradually to a minimum and increases again toward the end of the stroke. He states that the final pressure may be higher than the maximum at the start of the extrusion if the bottom of the part is very thin. Figure 2:8 is a plot of the fluctuation in pressure that he obtained in extruding from a $\frac{1}{4}$-inch blank a cup of aluminum alloy 2 inches in diameter with both wall and bottom $\frac{1}{32}$-inch thick. He did not describe the method used in determining this curve, but the work was done on a commercial press.

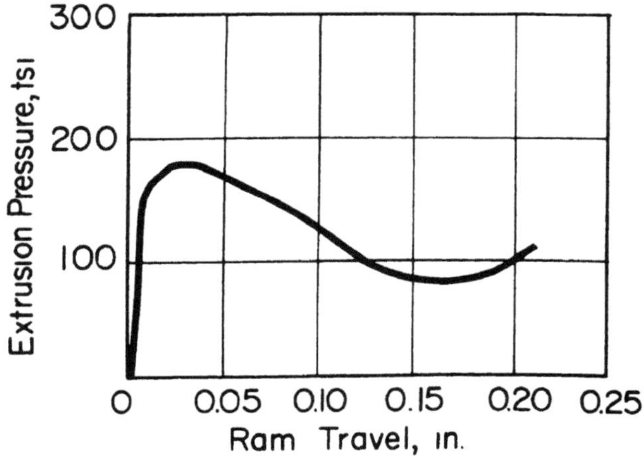

FIG. 2:8. FLUCTUATIONS IN PRESSURE DURING BACKWARD EXTRUSION OF AN ALUMINUM ALLOY CUP (*after* Boston[3])

The investigators just cited dealt with a limited number of metals. Recently, Ridings and Angus[24] investigated the factors that influence backward extrusion. As a result of studying the load-penetration diagrams of fourteen metals, they reached a tentative conclusion that will be highly significant if the results are confirmed. They found that the diagrams are of two types. In one, the pressure remains constant after reaching the maximum; in the other, the pressure falls rapidly from the maximum. They attributed the rapid drop in pressure to softening by heat and decided that materials showing this effect might, if preheated, require considerably less extrusion pressure. Conversely, those materials having constant pressure graphs should not be benefited by preheating the blank.

To test their theory, they made a limited number of experiments in which blanks were preheated to 1100°F, transferred rapidly to the die, and extruded with a reduction of area of 50 per cent. Among their data were:

Material	*Type of Diagram*	*Reduction in Pressure, %*
0.1%-carbon steel	Constant pressure	13
0.3%-carbon steel	Constant pressure	15
13%-chromium steel	Falling pressure	42
Titanium	Falling pressure	55

These results indicate that materials react differently when preheated before extrusion; and that, in some cases, moderate preheating may be beneficial. Apparently, the specific metal or alloy must be tested to determine the effect of preheating.

SPEED IN IMPACT EXTRUSION

The influence of speed in impact extrusion is probably the most controversial point in the process. Opinions range all the way from that of Gabler,[13] who states that in the backward extrusion of aluminum the impact action may be preferred only from the point of view of production rates, to that of Jevons,[14] who postulates that work hardening requires a small but finite interval of time to occur and it is probable that this fact accounts for some of the success of impact extrusion. Once again, recourse must be made to experimental investigations.

Von Zeerleder and von Burg[15] investigated the requirements of force related to speed, using a hydraulic press. Relatively slow speeds were employed and although this work does not approach commercial practice it does indicate speed effects. Using 99.5% aluminum, they found that increasing the rate of extrusion from a minimum (unspecified) rate to 0.008 inch a minute resulted in an increase in consumption of energy from 1560 to 2230 pounds per inch. An additional increase in rate to 1.2 inch a minute—maximum for the machine—resulted in an increase in consumption of energy to only 2950 pounds per inch. These investigators state that it would be necessary to assume an additional increase in consumption of energy of 20 to 30 per cent if the speed were increased to the normal extrusion rate, which they give as 8 to 16 inches a minute.

Derge and Stewart[16] also used a hydraulic press to study the extrusion of tin alloys containing varying proportions of copper up to 5 per cent. Again, these were low-speed tests; at a speed of 0.1 inch a minute, the force was 25,000 pounds. Extrapolation of their results to commercial speeds however, gave an indicated force of 37,000 pounds for a copper-tin alloy containing 0.25 per cent copper, in extruding a shell with a wall 0.010 inch thick. This compares with 39,000 pounds for a similar alloy with a wall of the same thickness, as determined on a commercial extrusion press by the author. It might be well to point out here that Herrmann[17] states that the impact velocities in the commercial extrusion of aluminum reach 210 feet a second.

The fact that extrapolation from slow-speed tests, although admittedly a rather questionable procedure, yields results that compare with those obtained on commercial equipment, is an indication that the slow-speed work is valuable in studying the forces involved. This point will be apparent in the discussion of work done on other materials.

In the backward extrusion of unalloyed aluminum, speed is not essential. It is possible to make a shell in a hydraulic press by starting and stopping repeatedly. The length of the shell will fall within the limits established for a similar part produced in uninterrupted extrusion. The major difference will be in the appearance of the two parts. The shell that has been made in a series of starting-and-stopping operations will show marks at each point where the extrusion edge of the punch stopped during the operation. The part produced conventionally will have smooth walls.

The results discussed here apply to extrusions of aluminum. Foisy[18] says attempts to forward-extrude copper tubes slowly instead of under impact forces result in breaking tools because of excessive pressures. He states that the success of the operation on copper is related to the rate of deformation.

These conclusions were reached many years ago and are not in accord with recent experience. Fukui and associates[19] included copper and brass in an extensive study of impact extrusion. Low-speed tests were run on a universal testing machine; high-speed tests were run on a mechanical press operating at 160 strokes a minute. They found only insignificant differences in the pressures required at the two speeds, and experienced no difficulties in making the parts.

Turning now to steel, Byles[20] reports that, although speeds of

10 to 30 inches a second are suggested frequently, many successful steel parts have been produced at much slower speeds, particularly in heavy hydraulic presses. He suggests that high speeds are particularly applicable to the production of plain cylinders with walls of constant thickness.

Wilson,[21] on the other hand, reports that within limits, higher rates of straining are beneficial in the forward extrusion of low-carbon steel. Lower extrusion pressures were required at average speeds of one inch a second than at one-tenth that speed. The differences depended on the reduction achieved. With a reduction of 90 per cent, the higher rate required 33 per cent less pressure than the slow rate; at 40 per cent reduction, there was no difference in the pressure requirements; at 20 per cent reduction, the higher rate required 20 per cent more pressure than the slow rate. He concluded that the difference was caused by the fact that the heat of deformation increased with the reduction.

The investigations mentioned are sufficient to indicate the differences of opinion expressed by various men who have studied the process. Most of those that consider the impact force either advantageous or essential did their work a number of years ago however. In most later investigations, it was concluded that impact is not significant, at least in so far as slow speed versus the speeds obtainable on commercial mechanical presses is concerned. Thus Gabler's statement, made about 25 years ago, that speed is advantageous only from the point of view of increased production, appears to be supported by later work. It is probable that tooling and lubrication are more important in this speed range than speed itself. These subjects will be discussed in later chapters.

On the other hand, work now in progress indicates that high-rate forming processes, such as explosive forming, can be used for the impact extrusion of metals that form only with difficulty at low rates. It appears from these initial results that high speed is advantageous in forming such materials, some of which appear to be relatively brittle at normal forming rates.

WORK OF DEFORMATION

Johnson[22] postulates that in impact extrusion, the potential energy of the ram is dissipated in:

1. Extrusion of the metal.
2. Deflection of the anvil under the impact.
3. Rebound of the ram.
4. Overcoming friction.
5. Stress waves in the tools that appear as heat or sound waves.

As a result of a series of experiments on the forward extrusion of aluminum, he concluded that about 90 per cent of the work expended during the process appears as heat, which is generated near the orifice of the die.

The work of deformation that is transformed into heat causes a rise of temperature in the shell and in the tools. This work of deformation begins only after plastic flow starts. It can be measured by integrating the area under the extrusion curve but it can also be approximated by calculation.

Siebel[23] states that the work done by deformation of a shell is converted into heat with the exception of a small percentage that remains as latent energy in the strengthened metal. For the temperature rise resulting from working, he gives the equation:

$$\Delta T = \frac{\varphi_h K_f}{wc\gamma}$$

where w = mechanical equivalent of heat
c = specific heat of the metal
γ = density of the metal
φ_h = deformation
K_f = deformation strength

Pineau[8] states that for a given metal the work necessary to extrude a part increases with the reduction of area and the thickness of the blank.

REFERENCES

1. Wagner, W. S., Plastic Working of Metals, *Trans ASME*, 78:401 (1956).
2. Templin, R. L., Effects of Cold Working on Physical Properties of Metals, *Trans AIME*, 89:466 (1930).
3. Boston, J. R., Cold Impact Extrusion of Aluminum Parts for Douglas Aircraft, *Machinery*, 51, July 1945:138.
4. Mabb, P., Impact Extrusion, *Machinery* (London), 76, February 9, 1950:192.

5. FELDMANN, H. D., Investigation on Power and Work Required in the Cold Forging of Various Steels, *Stahl und Eisen*, 73, January 29, 1953:165.
6. SABROFF, A. M., SANNICANDRO, R. A., and FROST, P. D., Design Considerations for Cold Extrusion of Titanium, TP 119, *American Society of Tool Engineers*, 1958.
7. HAUTTMANN, H. and PESSL, H. J., The Aluminum Bearing Non-Aging Steels and Their Cold Extrusion, *Steel Processing*, November 1950:557.
8. PINEAU, M., Backward Impact Extrusion, *La Pratique des Industries Mecaniques*, 37, June 1954:163.
9. SIEBER, K., Developments on the Cold-Flow Pressing and Extrusion of Steel, *Machinery* (London), May 22, 1952:891.
10. KÜHNE, J. H., *Load in Impact Extrusion of Hollow Cylindrical Parts*, Verlag W. Knapp, Halle, Germany, 1951.
11. GALLOWAY, D. F., Some Experiments on the Impact Extrusion of Aluminum, Copper, and Steel, *Machinery* (London), 93, August 13, 1958:356.
12. PHILLIPS, R. M., High Speed Photos Reveal Extrusion Rate Phenomenon, *Iron Age*, 162, December 2 1948:102.
13. GABLER, P., Precision Impact Extrusion and Die Casting of Aluminum in Heavy Industry, *Aluminium*, 25:194 (1943).
14. JEVONS, J. D., *The Metallurgy of Deep Drawing and Pressing*, p. 109; Chapman Hall, London, 1940.
15. VON ZEERLEDER, A. and VON BURG. K., Relations of Power Demands in Extrusion of Pure Aluminum and Various Aluminum Alloys, *Schweizer Archiv*, 3:109 (1937).
16. DERGE, G. and STEWART, J. W., Extrusion of Tin and Tin Alloys, *Trans AIME*, 137:389 (1940).
17. HERRMANN, E., Impact Extrusion of Aluminum, *TZ fur Praktische Metallbearbeitung*, 47, July 1937:528.
18. FOISY, G. A., The Manufacture and Application of Extruded Copper Tubes, *Trans. AIME, Institute of Metals Division*, 1927, 93.
19. FUKUI, S., KUDO, H., and SEINO, J., Study on Impact Extrusion Method for Non-Ferrous Metals, *Report of Inst. Science and Technology, Univ. Tokyo*, 11:91 (1957).
20. BYLES, H. W., Impact Extrusion, *Metal Treatment*, March 1958: 107.
21. WILSON, D. V., Metallurgical Requirements of Steels for Cold Extrusion, *Sheet Metal Industries*, 30, June 1953:513.
22. JOHNSON, W., Research into Some Metal Forming and Shaping Operations, *Sheet Metal Industries*, 34, February 1957:121.
23. SIEBEL, E., The Plastic Forming of Metals, *Steel*, November 13, 1933:26.
24. RIDINGS, J. B. and ANGUS, N. S., Experiments on the Reverse Impact Extrusion of Certain Metals, *J. Institute of Metals*, 90, December 1961:107.

Chapter 3

EXTRUSION PRESSURE

A number of factors influence the pressure required for extrusion and they are interrelated. During the course of numerous investigations, the effects of changes in individual factors have been determined, and it is possible to discuss some of them singly.

Among the major factors influencing the pressure are the metal and its properties, the size and shape of the part, the thickness of the wall to be produced—or conversely the reduction of area—the temperature, and the effect of lubrication. The last will be discussed in a succeeding chapter. Among minor factors are the thickness and shape of the blank, and the size and orientation of the grain.

EFFECT OF THE METAL AND ITS PROPERTIES

In the present state of the art, it is not possible to separate the effect of composition from the effect of mechanical properties in a single alloy system.

For example, in the backward extrusion of three grades of aluminum under the same conditions, the extrusion pressure increased in the following proportions:

Aluminum, %	*Ratio using* 99.93% *as the base* Extrusion Pressure	Hardness
99.93	1	1
99.85	1.1	1.7
99.5	1.2	2.6

The 99.93% grade is considerably softer than the other grades but the extrusion pressure does not appear to reflect this difference.

Chopin and Chopin[1] state that the extrusion pressure for a given diameter is inversely proportional to the Brinell hardness in the backward extrusion of aluminum and aluminum alloys. Their data, based on commercial aluminum appear in Table 3:1.

TABLE 3:1

RELATIONSHIP BETWEEN HARDNESS AND EXTRUSION PRESSURE FOR ALUMINUM ALLOYS*

Alloy	*Brinell Hardness*	*Ratio of Pressure to that Required for A5*
A5	19	1
A-SG	35	1.29
A-G3	61	1.95
A-Z8GU	60	1.96
A-G5	71	2.34

* Chopin and Chopin[1]

The same relationship did not hold for copper, tin, and lead. For example, they found that the extrusion pressure for copper (Brinell hardness 36) was 2.32 times that required for commercial aluminum; if their relationship in hardness were valid for copper, the ratio should have been 1.3.

Gokyû *et al.*[2] appear to confirm the Chopins' results. They determined that it required only 80 per cent as much force to extrude annealed aluminum as it did to extrude as-rolled material from the same lot. Here, hardness appears to be the controlling factor.

Pineau[3] states that, although the extrusion pressure for aluminum and copper is inversely proportional to the hardness, the function is not linear. On the other hand, he found a linear relationship between extrusion pressure and rupture strength for these metals. Lead did not conform to the relationship found for aluminum and copper.

The failure of lead, and probably tin, to follow the relationships found for the harder metals is not surprising because these two metals are being hot-worked during extrusion whereas the harder metals are undergoing cold work.

Sieber[4] states that the pressure required for extrusion of steel, and the strength of the finished part, increase with the content of carbon, manganese, and silicon. This statement appears to be confirmed by the work of Hauttmann and Pessl,[5] based on an investigation of eleven steels of varying carbon content. They found that the extrusion pressure increased with the carbon content. They stated also that the extrusion pressure was proportional to the tensile strength.

TABLE 3:2

RATIO OF EXTRUSION PRESSURES FOR VARIOUS MATERIALS*

Material	*Ratio of Extrusion Pressure* (Lead = 1)
Lead	1
Tin	1.2
1100 Aluminum	2.2
6061 Aluminum	4.3
7075 Aluminum	8.6
Copper	4.9
C1010 steel	12.6

* Various sources

Lloyd and Kopecki[6] reported that it required approximately 20 per cent greater pressure to forward-extrude a shell from AISI 4615 steel (0.15% C, 0.60% Mn, 1.75% Ni, 0.20% Mo, 0.10% Cr) than to extrude an identical shell from 0.15% plain

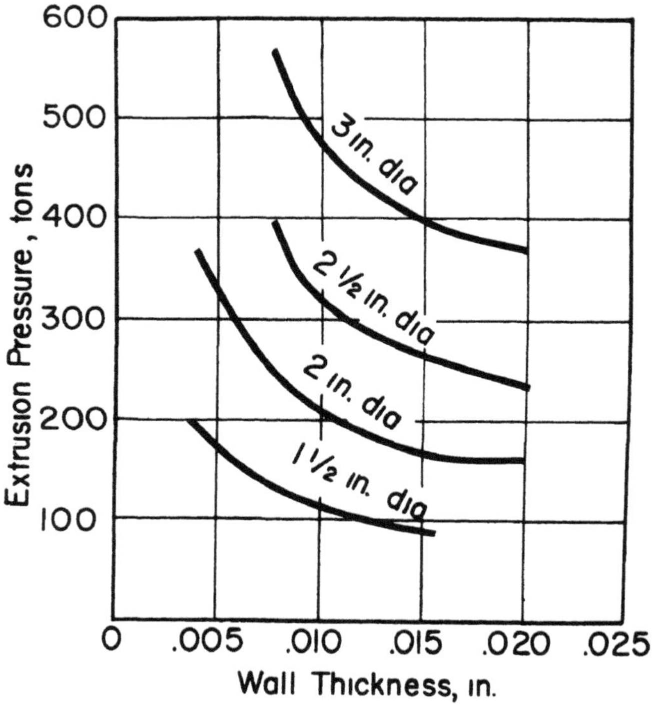

FIG. 3:1. EFFECT OF DIAMETER ON PRESSURE REQUIRED TO EXTRUDE SHELLS HAVING WALLS OF DIFFERENT THICKNESS (Lang, Ansell and Wilson[17])

carbon steel. Similarly, a steel of 0.20% C and 0.75% Cr required about 22 per cent more pressure than a steel of 0.20% carbon.

There is a wide variation in requirements for extrusion pressure from one alloy system to another. An indication of the differences is given in Table 3:2, assembled from various sources, for materials ranging from lead to low-carbon steel.

EFFECT OF SIZE AND SHAPE OF PART

The extrusion pressure is directly proportional to the cross section of the part. Figure 3:1 indicates the effect of changes in diameter on the pressure required to extrude shells having walls of various thicknesses.

Although there appears to be no published information on the effect of shape on extrusion pressure, a shape having a circular cross section and smooth walls offers the least resistance to flow and therefore it should require the minimum extrusion pressure. Use of the proper radii with any shape will lead to the minimum pressure for that shape. This matter will be discussed further in the chapter on design of parts.

EFFECT OF WALL-THICKNESS

The thickness of the wall, for a given diameter, is of major importance in impact extrusion. It determines the deformation involved and the severity of the operation. The effect of wall-thickness, therefore, has been the subject of numerous investigations.

Kühner[7] was one of the earliest to note that the pressure required for a given cross-sectional area is inversely proportional to the thickness of the wall. Chopin[8] noted that the required pressure is not a linear function of the wall-thickness but that it rises rapidly with very thin walls.

The author determined that for a relatively limited range of thin walls (between 0.003 and 0.0115 inch), the relationship between total pressure and wall-thickness was exponential and could be expressed as

$$P = \frac{A}{x^n}$$

where P = pressure in tons
x = wall-thickness in thousandths of an inch
A and n are constants

The relationship was found to be valid for aluminum, lead, and tin for round shells of small diameter.

The effect of wall-thickness on the extrusion pressure for parts backward-extruded from commercial aluminum is indicated in Figure 3:1. For example, extrusion of a round shell, two inches in diameter, required about 160 tons at a wall-thickness of 0.020-inch, and 250 tons at a wall-thickness of 0.0075-inch. Conversely, at a wall-thickness of 0.015-inch, pressure increases from 90 tons for a 1.5-inch diameter shell to 400 tons for a 3-inch diameter shell.

Shoemaker[9] reported that in the backward extrusion of Type 1100 aluminum, the required pressure was 30,000 *psi* at a wall-thickness of 0.18-inch and increased to 70,000 *psi* at a wall-thickness of 0.04-inch.

Byles[10] states that reducing the wall-thickness from 0.11-inch to 0.011-inch increases the extrusion pressure for copper from 121,000 to 262,000 *psi*.

Hauttmann and Pessl[5] discussed the effect of wall-thickness in the extrusion of a number of carbon steels but reported their data in a more logical manner based on reduction of area. Their work will be discussed in the following section.

EFFECT OF REDUCTION OF AREA

Basing the extrusion pressure on the reduction of area has become rather common and this method is used widely in determining requirements for commercial operations.

In backward extrusion, the reduction of area can be expressed as:

$$R = 100\left(1 - \frac{a}{A}\right)$$

where R = reduction of area in per cent
a = area of extruded part
A = area of the blank (or cavity of the die)

For parts having circular cross-sections, the formula becomes:

$$R = \frac{d^2}{D^2}$$

where d = diameter of punch
D = diameter of die

An advantage of impact extrusion is the possibility of making large reductions of area in a single step. For example, Howard[11] states that reductions up to 98 per cent are possible with aluminum and aluminum alloys; up to 90 per cent with copper; up to 75 per cent with copper alloys; and up to 80 per cent with nickel.

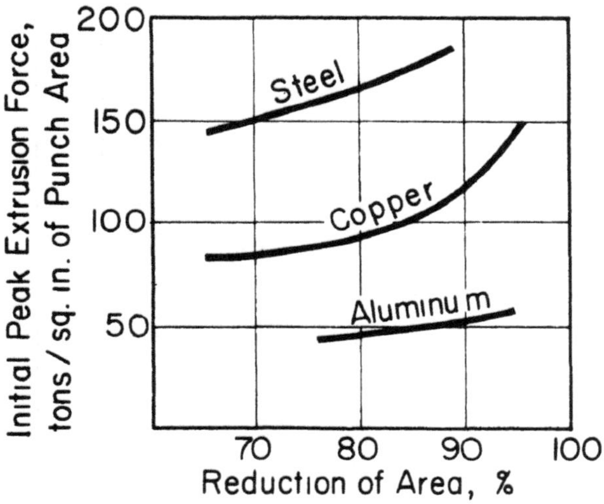

FIG. 3:2. RELATIONSHIP BETWEEN EXTRUSION PRESSURE AND REDUCTION OF AREA IN BACKWARD EXTRUSION OF STEEL, COPPER, AND ALUMINUM (Galloway[12])

Galloway[12] remarks that the practical limit of reduction in backward extrusion can be determined by expressing the peak load in terms of load to unit area of the punch. Setting his limits at a load of 150 long tons to the square inch, his maximum reductions are 95 per cent for copper and 69 per cent for low-carbon steel. His data on the relationship between extrusion pressure and reduction of area are shown in Figure 3:2.

Byles[10] reports that in backward extrusion the maximum reduction of area in commercial practice varies with the metal; with aluminum, reductions of 99 per cent are possible; with steel, reductions between 55 and 70 per cent are practical. He notes that greater reductions have been achieved with steel, experimentally, but have not been maintained in actual production.

Reduction of area in forward extrusion can be obtained by

formulas similar to those used for backward extrusion. As shaped blanks are used, however, the simple relationship shown for round shells cannot be applied.

Hauttmann and Pessl[5] determined the effect of reduction of area in forward extrusion for a series of carbon steels ranging

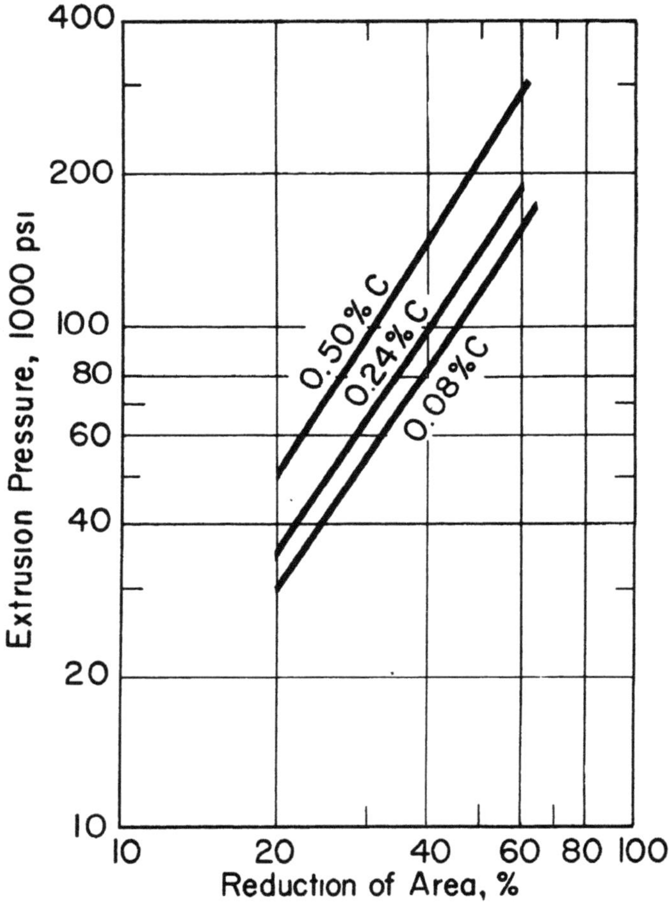

FIG. 3:3. RELATIONSHIP BETWEEN EXTRUSION PRESSURE AND REDUCTION OF AREA IN FORWARD EXTRUSION OF STEELS (Hauttmann and Pessl[5])

from 0.08 to 0.50% carbon. They found that a direct relationship existed between reduction of area and extrusion pressure in the range of 20 to 75 per cent reduction. Some of their data have been plotted in Figure 3:3 and the relationship appears to be exponential.

Feldmann[13] investigated both backward and forward extrusion on a series of steels. His work offers an opportunity to compare the relative extrusion pressures required for the two types. Data on one of his steels have been plotted in Figure 3:4. These results show that much higher pressures are required for backward extrusion than are necessary for forward extrusion, as should be expected because greater friction is encountered in the former. For example, it requires as much pressure to obtain a 20 per cent reduction of area by backward extrusion as it

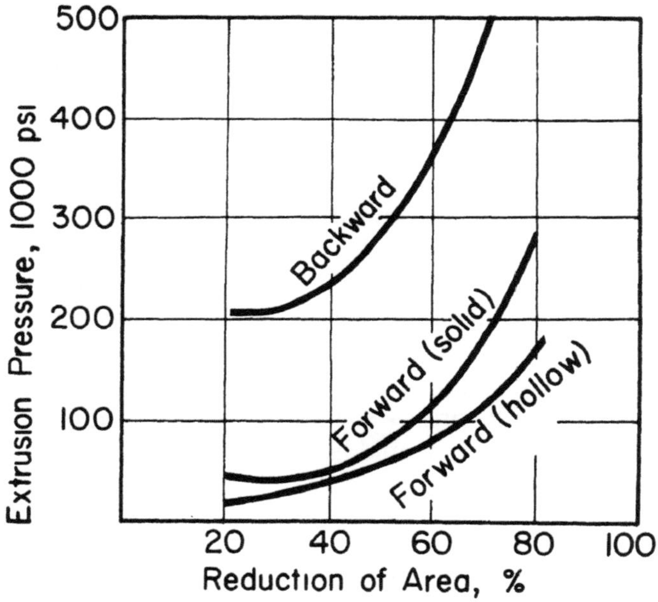

FIG. 3:4. COMPARISON OF EXTRUSION PRESSURES REQUIRED FOR BACKWARD AND FORWARD EXTRUSION OF STEEL (Feldmann[13])

does to obtain a 70 per cent reduction by forward extrusion. It is interesting that forward extrusion of solid parts requires more pressure than forward extrusion of hollow pieces.

EFFECT OF TEMPERATURE

As there is a definite relationship between the strength or hardness of a material, according to some observers, and the required pressure, it should be expected that anything that reduces the hardness should also reduce the extrusion pressure. Temperature is such a factor.

Pineau[3] states that the maximum extrusion pressure decreases linearly with rising temperature. Von Zeerleder and von Berg[14] agree. These men determined the pressure required for the backward extrusion of aluminum and several aluminum alloys. Setting 100 per cent at 68 F as the criterion for the required pressure, they found the following relationship for aluminum:

68 F	300 F	390 F	480 F
100%	59%	41%	37%

These investigators found a similar relationship during the extrusion of a number of aluminum alloys. However, the work was done on a low-speed press and does not approach commercial operating conditions.

Galloway[12] found that the extrusion pressure rises to the maximum initially and there is a second, but lower peak, at the maximum penetration of the punch. During the extrusion of copper blanks at temperatures up to 480 F, he determined that the maximum pressures decreased linearly with rising temperature; in his experiments the pressure dropped from 125 tons to the square inch at 32 F to 100 tons to the square inch at 480 F. However, the pressure at maximum penetration of punch remained constant. Since the maximum pressure reached the same value as the pressure at maximum penetration at 275 F, he concluded that there would be no advantage in exceeding this temperature.

TABLE 3:3

EFFECT OF PREHEATING BLANKS ON EXTRUSION PRESSURE FOR ALUMINUM ALLOYS*

Alloy	*Extrusion Pressure, tsi, after Preheating to Temperature Indicated*		
	68 F	250 F	800 F
1100-0	52	35	8
6061-0	70	50	25
2024-0	125	100	50
7075-0	175	120	65

* Boston[15]

Boston[15] also reported a series of tests on aluminum and aluminum alloys. Unlike some of the other work, his was performed on regular production equipment. His data, given in Table 3:3, indicate the significant effect of temperature on the required pressure. It should be noted that in commercial operations, the temperature of the tools will rise and this should influence the pressure requirements. When producing parts that require pressures near the maximum permitted on the punch, heating the blanks before extrusion, may be the difference between success and failure in producing a part.

That reducing the extrusion pressure by heating the blanks is not applicable to all materials is indicated by the work on load-penetration diagrams reported by Ridings and Angus (see Chapter 2, page 24).

EFFECT OF THICKNESS OF BLANK

Pineau[3] states that the maximum extrusion pressure in backward extrusion increases with the thickness of the blank. He notes also that it is always possible to obtain a thick base but, below a certain thickness, it is impossible to extrude the metal remaining between punch and die.

In working with aluminum, Galloway[12] found that increasing the blank thickness from $\frac{1}{16}$ to $\frac{1}{4}$ inch resulted in an increase in maximum extrusion pressure from 32 to 47 tons to the square inch. There was no additional increase in pressure up to a thickness of $\frac{3}{8}$ inch, the maximum he employed. On the other hand, during the extrusion of steel with reductions of area of 63 and 85 per cent, the extrusion pressure remained practically unchanged for blank thicknesses varying from $\frac{1}{4}$ to $\frac{9}{16}$ inch.

The author found that increasing the blank thickness from 0.1 to 0.25 inch caused an increase in pressure of about 50 per cent during the extrusion of antimonial lead; increase in blank thickness from 0.06 to 0.23 inch when extruding commercial aluminum caused an increase of about 30 per cent in the extrusion pressure.

SHAPE OF BLANK

Blanks are generally shaped to the bottom contour of the finished part for backward extrusion. Herrmann[16] states, however, that it is unnecessary to use such shapes; and that hexagonal or even square blanks can be used for the production of circular shells. It should be noted that he was discussing the extrusion of relatively soft aluminum. Others have found that use of a square or hexagonal blank increases the extrusion pressure unless the corners have been rounded.

Lang, Ansell, and Wilson[17] suggest that if the blanks are preformed into a conical shape and fed into the die with the apex of the blank toward the opening of the die, the pressure during backward extrusion will be considerably reduced.

In forward extrusion, the blank is generally much thicker than that used in backward extrusion. Usually, it is preformed into a thick-walled cup, which, depending on the part that is to be produced, may or may not have a hole in the bottom. No data have been found on the effects of the shape of the blank on extrusion pressure during forward extrusion.

SIZE OF GRAIN

Little information has been published on the effect of the size and orientation of grain on extrusion pressure and it is somewhat contradictory.

Stelljes[18] states that fine-grained, fully-annealed material is most satisfactory for extrusion. In the backward extrusion of aluminum, the writer found no difference in ease of extrusion or pressure until extremely coarse-grained material was employed. With this, there was about five per cent reduction in extrusion pressure.

Lang, Ansell, and Wilson[17] report that for aluminum, a coarse grain structure with its associated low yield strength requires the lowest extrusion pressure. They suggest that all sizes of grain up to 0.04 inch might be used if the resulting surface is sufficiently smooth.

There is some indication that extrusion pressures are slightly lower with a blank cut off a rod, in which the fibers are parallel to the vertical axis of extrusion, than with those blanked from strip, in which the fibers are at right angles to the axis. Herrmann[16] claims that cast blanks are easier to extrude than rolled blanks.

FORMULAS FOR EXTRUSION PRESSURE

The preceding discussion has involved individual factors that affect the pressure required to extrude various materials. In practice, these factors are interrelated and their effects must be considered together in determining the pressure required to produce a given part.

There have been numerous attempts to devise formulas for

determining the required pressure. Development of Sieber's formula was discussed in Chapter 2 as an illustration of the methods employed.

Sachs[19] has proposed a simple equation for the extrusion pressure. His formula is based on the dimensional change in the part and the compressive resistance of the metal.

His equation is:

$$P = Ak \ln \frac{A}{a}$$

where A = cross-sectional area of the die recess
a = cross-sectional area of the shell
k = constant for the metal

He states that the value of k depends on the actual temperature of the partially extruded blank and is approximately equal to twice the compressive resistance.

The formulas of Sieber and Sachs do not take the internal heating effects, resulting from working, into account directly, although Sachs attempts to compensate for them in his value for the constant, k. Recently, Pineau[3] proposed another formula which does consider temperature changes. He suggests that the formula takes the form:

$$P = A\left[f\left(\frac{A}{a}, e\right)\right] \times \frac{3.6\, R_r - 16}{20} \times \frac{\theta_0 - \theta}{\theta_0 - 20}$$

where e = thickness of blank, millimeters
A = cross-sectional area of die recess, square centimeters
a = cross-sectional area of shell, square centimeters
R_r = rupture strength (kg/mm^2) of metal being extruded
θ = actual extrusion temperature (°C)
θ_0 = temperature corresponding to minimum pressure (varies with the metal)

He states that for aluminum and copper, the pressure is related directly to the rupture strength and the formula becomes:

$$P = A[3.6\, R_r - 16]$$

Pineau states that the first formula given is too complicated to be practical but he prepared a nomograph that shows the relationships. Of interest here, is the fact that he calculated the extrusion pressures for several materials by Sachs' formula and his own, and compared the values with those obtained experimentally. He calculated similar values with one of Sieber's formulas, also, but unfortunately chose an equation for piercing instead of one for extrusion. Using the same symbols as those used in Sachs' equation, the Sieber equation for piercing is:

$$P = Ak \times \tfrac{1}{2}\left[\ln\frac{A}{a} + \ln\frac{A}{A-a} + \ln\frac{A-a}{a}\right]$$

Table 3:4 shows some of the results obtained by Pineau. Although the formulas of Sachs and of Sieber approximate the measured values, they are somewhat erratic. Pineau's nomograph is consistently closer to the actual values, particularly when the temperature is considered.

TABLE 3:4

COMPARISON OF EXTRUSION FORMULAS*

Material	*Ratio*	*Temperature,* F	*Extrusion pressure, long tons* Measured	Pineau Formula	Sachs Formula	Sieber Formula
99.5% aluminum	3	68	13.6	13.2	17.4	12.7
(Brinell 32)	10	68	25.5	27.4	16.0	26.6
	30	410	23.8	23.3	39.4	45.1
Copper	3	68	36.0	38.6	27.7	38.1
(Brinell 51)	6	320	43.7	46.7	45.2	57.4

* Pineau[3]

Fukui and associates[20] also prepared a formula for the maximum pressure required in the backward extrusion of a cylindrical shell. The formula is:

$$P_{\max} = KF_1\left(\frac{C}{D_P}\right) \times F_2\left(\frac{T}{D_P}\right) \times Y_m$$

where K = a constant

F_1 = ratio of extrusion clearance to punch diameter

F_2 = ratio of blank thickness to punch diameter

Y_m = mean flow stress of the metal (determined experimentally)

D_p = punch diameter

C = clearance

T = thickness of blank

Based on experimental work by static extrusion, this formula becomes:

$$P_{\max} = 1.60\left(\frac{C}{D_p}\right)^{-\frac{1}{3}} \times \left(\frac{T}{D_p}\right)^{\frac{1}{6}} \times Y_{0.55}$$

where $\frac{C}{D_p} = 0.015$ to 0.100

$\frac{T}{D_p} = 0.2$ to 1.0

A comparison of extrusion pressures determined by Fukui *et al.*, with actual pressures is given in Table 3:5 for several metals. These investigators stated that they compared their results with those obtained by using Pessl's nomograph (published by Everhart[21]) and found that the latter were less accurate. They gave no direct comparisons.

TABLE 3:5

COMPARISON OF ACTUAL WITH CALCULATED EXTRUSION PRESSURES*

Material	*Hardness of Blank* (Vickers)	*Mean Flow Stress* kg/mm²	*Punch Diameter* mm	T/D_p	C/D_p	*Maximum Pressure* Measured kg/mm²	tsi	Calculated kg/mm²	tsi
Aluminum (soft)	20	12.7	12.1	0.38	0.015	69	44	70	44.5
			19.0	0.21	0.081	34	21.6	36	22.8
Aluminum (hard)	43	16.2	12.7	0.39	0.031	72	45.7	70	44.5
			11.25	0.89	0.063	57	36.2	55	35
Duralumin	65	35.8	12.7	0.39	0.031	147	93.5	154	98
			11.25	0.44	0.100	106	67.5	108	68.8
Copper (soft)	38	35.0	13.1	0.38	0.015	208	132	193	123
			12.0	0.67	0.063	140	88	132	83.6
Copper (hard)	110	—	12.7	0.38	0.031	182	116	180	114
			11.25	0.89	0.100	141	89	143	92

* Fukui, Kudo, and Seino[20]

Fukui and his associates state that the factors affecting high-speed extrusion are somewhat more complicated than those involved in static extrusion. They found that the maximum dynamic extrusion pressures for metals of high melting point were somewhat lower than pressures for static extrusion, whereas their flow stresses were somewhat higher. They concluded, however, that the maximum extrusion pressures for these metals at normal rates of extrusion on mechanical presses may be safely regarded as equal to those in static extrusion.

Any of these formulas or the nomographs prepared from them should be used with caution merely for a rough estimate of the pressure requirements. An important factor ignored in all of them is lubrication, which can be highly influential. For example, the author was able to vary the extrusion pressure by 30 per cent in the backward extrusion of aluminum merely by changing the lubricant. The subject of lubrication is discussed in another chapter.

REFERENCES

1. Chopin, R. and Chopin J., Impact Extrusion of Light Metals, *Revue de l'Aluminium*, 32:1150 (December 1955).
2. Gokyu, I., Suzuki H., and Shinji, K., Studies on Impact Extrusion. VII: On Cold Extrusion Forces, *Nippon Kinzoku Gakki-Si*, 19:351 (June 1955).
3. Pineau, M., Backward Extrusion, *La Pratique des Industries Mecaniques*, June 1954:163.
4. Sieber, K., Developments on the Cold-Flow Pressing and Extrusion of Steel, *Machinery* (London), May 22, 1952:891.
5. Hauttmann, H. and Pessl, H. J., The Aluminum Bearing Non-Aging Steels and Their Cold Extrusion, *Steel Processing*, November 1950:557.
6. Lloyd, T. E. and Kopecki, E. S., Cold Extrusion of Steel, *Iron Age*, 164, May 4, 1949:90.
7. Kuhner, O., The Manufacture of Tubular Aluminum Articles by Extrusion, *Metal Industry* (London), 47:535 (1935).
8. Chopin, R., Researches on the Technique of Impact Extrusion, *Revue de l'Aluminium*, 33:365 (April 1956).
9. Shoemaker, J. D., Impact Extrusion of High Strength Aluminum, *American Machinist*, July 6, 1953:129.
10. Byles, H. W., Impact Extrusion, *Metal Treatment*, March 1958:107.
11. Howard, F., Designing with Impact Extruded Components, *Engineering Materials and Design*, October, 1958:26.
12. Galloway, D. F., Some Experiments on the Impact Extrusion of Aluminum, Copper and Steel, *Machinery* (London), 93:356 (1958).
13. Feldmann, H. D., Investigations on Power and Work Required in Cold Forging of Various Steels, *Stahl and Eisen*, 73, January 29, 1953:165.
14. Von Zeerleder, A. and von Berg, E., Relationships of the Force Requirement in the Extrusion of Tubes of Pure Aluminum and Various Aluminum Alloys, *Schweizer Archiv*, 3:109 (1937).
15. Boston, J. R., Extrusion of Aluminum Parts for Douglas Aircraft, *Machinery* (N.Y.), July 1945:139

16. HERRMANN, E., The Cold Extrusion of Aluminum, *TZ für Praktische Metallbearbeitung*, 47:528 (July 1937).
17. LANG, R. F., ANSELL, W. R. and WILSON, D. V., Impact Extrusion: German Practice, 1946, *B.I.O.S. Final Report* No. 1405, 1946.
18. STELLJES, H. A. J., On the question of Raw Materials and Their Properties in the Impact Extrusion of Light Metals, *Metallforschung*, 2:341 (1947).
19. SACHS, G., A Review of Impact Extrusion and Some Related Processes, *Metal Industry*, 28:533 (June 1951).
20. FUKUI, S., KUDO, H., and SEINO, J., Study on Impact Extrusion Method for Nonferrous Metals, *Report Inst. Science and Technology, University, Tokyo*, 11:81 (1957).
21. EVERHART, J. L., Impact (Cold) Extruded Parts, *Materials and Methods*, August 1953:111.

Chapter 4

TOOLING FOR IMPACT EXTRUSION

This chapter and the next two deal with topics that, taken together, determine the success or failure of the process of impact extrusion as a commercial operation. The topics are tooling, tool materials, and lubrication. Failure to use suitable tooling with adequate tool materials and proper lubrication can cause tools to fail in so short a time as to make the operation impractical.

Although this combination of factors has caused problems from the beginning of impact extrusion, they became more critical when attempts were made to extrude high-strength metals. The Germans produced steel parts by cold extrusion during World War II and their methods were seized by the Allies at the end of the war. In spite of this start, so many problems were encountered that it was several years before commercial production of steel parts was possible. The process is just beginning to reach its potential in many fields.

It is relatively easy to produce an experimental part by backward or forward extrusion but placing the operation on a commercial scale is not so simple. It requires the proper combination of tooling, tool material, and lubricant. For example, a lubricant that is perfectly satisfactory for lead may be useless for extruding aluminum. It cannot be emphasized too strongly that the three topics are interrelated, although they are discussed in separate chapters for convenience.

In impact or cold extrusion, the metal is forced to flow under compressive stresses. Tools must be designed to form orifices through which the metal flows, to direct the flow, and to assure a continuous supply of metal to the orifice. This requires the orifice to remain unchoked for the complete operation. It is desirable that the part be produced as rapidly as possible and that friction be minimized. As the tools required for backward extrusion differ somewhat from those used for forward extrusion, the two types will be discussed separately.

BACKWARD EXTRUSION

Gardner[1] states that for the backward extrusion of high-strength materials, the orifices of the tools should be short and create no effect of a funnel. The result of such an effect is unequal flow, choking, increased friction, and thinning of the lubricant. These remarks can be applied to the backward extrusion of any of the metals.

Designing the Punch

In backward extrusion, compressive stresses act on the punch whereas tensile stresses act on the die. Although the die can be supported, as will be discussed later, the punch must carry the full compressive load. Hence the ratio of slenderness is an important factor in the design of a punch because of the tendency to buckle under the high stresses involved. This ratio sets a limit on the length of the part that can be produced by backward extrusion.

Howard[2] suggests the following length/diameter ratios for practical use:

Lead and tin	6:8 to 1
Aluminum and its Alloys	5:7 to 1
Copper and its Alloys	3:5 to 1
Magnesium and its Alloys	6:7 to 1
Zinc and its Alloys	4:6 to 1
Nickel	3:4 to 1

These ratios appear to be somewhat conservative. For example, some producers recommend a maximum ratio of length to diameter of 8:10 to 1 for commercial aluminum and the softer aluminum alloys, although they agree on the 5 to 1 ratio for the high-strength alloys. A 6:9 to 1 ratio has been suggested as a limitation for magnesium alloys.

For steels, Howard and his associates[3] set a practical ratio of 4 to 1, whereas a producer recommended 3½ to 1 for low-carbon steels and 2½ to 1 for low-alloy steels.

The shape of the nose of the punch and its effect on backward extrusion have been investigated by Howard and his associates[3]. They used punches with flat, tapered, conical, and round noses to extrude steel. At reductions of area of 40 to 60 per cent, the extrusion pressure was lowest with a round-end punch, next with a conical-end, and highest with flat and tapered ends. At 80 per cent reduction, however, the extrusion pressure was practically independent of the shape of the nose. With this high reduction of area, they found that conical and rounded punches had a tendency to penetrate the phosphate coating, particularly toward the end of the stroke, and that scoring occurred. They concluded that these shapes should be used only for very shallow extrusions. As will be pointed out in subsequent paragraphs, the flat-bottomed punch with beveled edges is preferred by most of those that have discussed the shapes of the tools.

TABLE 4:1

EFFECT ON EXTRUSION PRESSURE OF RELIEVING PUNCH IN BACKWARD EXTRUSION*

Material	*Extrusion Pressure, psi*		*Ratio of B to A*
	Punch with Straight Sides (A)	Punch Relieved Behind Edge[a] (B)	
99.99% Al	115,000	55,000	0.47
99.5% Al	162,000	74,000	0.45
Al-Mn	170,000	88,000	0.52
Al-Mg-Si	216,000	98,000	0.46
Al-Cu-Mg	317,000	183,000	0.57
Al-Mg-Mn	317,000	171,000	0.54
Al-Mg	427,000	253,000	0.59

[a] Extrusion edge 10 ϕ, shank 9 ϕ
* Von Zeerleder and von Berg

Friction can have a marked effect on the extrusion pressure. To investigate the pressure required to extrude aluminum alloys, Von Zeerleder and von Berg[4] used a straight-sided punch and one having a reduced shank behind the extrusion edge. Table 4:1 shows that for the alloys studied relieving the punch resulted in a reduction of approximately 50 per cent in the extrusion pressure. It is general practice, whenever possible, to form the punch with a narrow extrusion edge and reduce the shank behind this edge, but there are occasions when this practice is undesirable. Morgan[5] points out that the use of an extrusion edge may cause the component to be extracted on the punch and require the use of a stripper. This may require increasing the length of the punch beyond the safe ratio of length to diameter.

The punch should be so designed that equal quantities of metal will flow at uniform rates on opposite sides. Under these

TABLE 4:2

SUGGESTED DIMENSIONS OF PUNCH FOR BACKWARD EXTRUSION OF CUPS

Material	*Diameter of Punch*	*Diameter of Shank* (D_1) inch	*Length of Extrusion Edge* (L) inch	*Angle of Bevel* (α) deg	*Source*
Aluminum	D	D – 0.012	[a]	25-27	Lang *et al*[6]
	D	D – 0.006	0.12/0.24	1/2[b]	Fischer[7]
	D	D – 0.004	0.06	[c]	Gabler[8]
	D	—	—	15	Boston[9]
	D	D – 0.020	0.010	15	Galloway[10]
Magnesium	D	D – 0.004	[a]	—	Dow[11]
Steel	D	D – 0.004	0.040	[e]	Morgan[5]
	D	D – 0.006	0.1875	5-7[f]	Leland[12]
	D	D – 0.006	0.12/0.24	5-7[b]	Fischer[7]
Zinc	D	D – 0.004	[a]	[d]	Lang *et al*[6]

[a] Sharp edge rounded to a radius of 0.008 inch
[b] Radius at edge 0.020 to 0.40 inch
[c] Radius into extrusion edge
[d] Bevel greater than that used for aluminum
[e] Flat with small radius into sidewall
[f] Face angle to center; no flat bottom

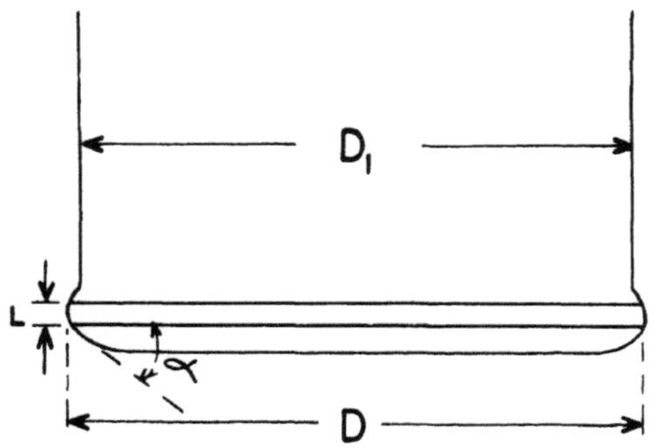

FIG. 4:1. TO ILLUSTRATE DIMENSIONS GIVEN IN TABLE 4:2.

conditions, bending stresses will be minimized. Failure to observe these precautions may result in eccentric products, broken tools or short life of a tool.

Dimensions of the punch and slope of its tip are altered somewhat for different materials. Table 4:2 lists dimensions recommended in various publications (see also Fig. 4:7). For most materials, a reduction of about 0.004 inch in the diameter of the shank appears to be selected. The length of the extrusion edge is varied from sharp with a small radius to a maximum of 0.24 inch.

Recommendations for the bottom of the punch are varied. Suggested shapes are (1) a round base, (2) a flat face with a bevel at the edge, and (3) a conical face with no flat area. Galloway[10] states that for the extrusion of aluminum the punch should either be rounded or have the greatest possible taper. Gabler[8] suggests the use of a flat center on the bottom of the punch and roughening this portion of the face to hold back surface metal and lubricant. Galloway[13] agrees that the bottom surface of the punch should be roughened and recommends the form of a grid. He notes that this design increases the extrusion pressure on shells with a very thin bottom but states that there are many applications where it is necessary to maintain stability of the punch at the beginning of a heavy reduction operation.

Leland[12] suggests a tapered punch with a 5 to 7 degree bevel for the extrusion of steel because with this shape the flow of the surface metal is distributed in such a manner that some lubricant remains at the bottom of the die cavity at the end of the stroke. Flatter punches increase the extrusion pressure and restrict the flow of the surface metal thus preventing feeding of lubricant to the side walls. Sharper punches reduce the extrusion pressure but cause the surface metal to flow in the early stages of the extrusion, thus robbing the punch of adequate lubricant later in the stroke.

There is general agreement that all points over which metal flows, particularly the bevel and the extrusion edge, should be highly polished. The punch should be removed from the press and repolished periodically. If this is not done, the walls of the extruded part may become scored and this would cause an increase in pressure.

Among other considerations, Morgan[5] notes that in the

extrusion of steel, the end of the punch will expand and the shank will take a permanent set causing a bulge at a distance of one-third to one-half of the length measured upward from the tip. If calculations for this bulging are not included in the design of the tools, the effect will combine with the recovery in the die (to be discussed later) and cause choking.

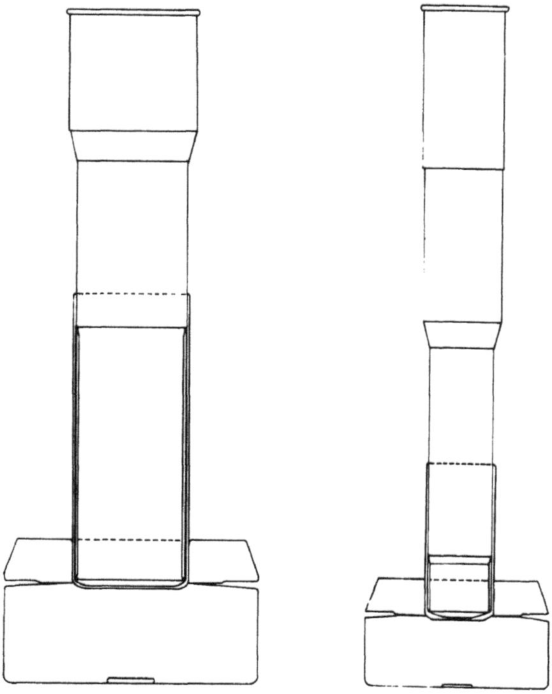

FIG. 4:2. TOOLS USED FOR BACKWARD EXTRUSION
(*a*) for Aluminum; (*b*) for Zinc
(Lang *et al*[6])

To reduce surface fatigue, the tools should be stress-relieved at regular intervals. The length of the production run, before relieving stress depends on the metal being extruded; the intervals should be shorter with high-strength alloys, such as steel, than with commercial aluminum, for example.

Some of the types of tools that have been suggested for backward extrusion are shown in Figures 4:1 to 4:4. Figure 4:2*a* is an example of a punch-and-die combination used in Germany for the production of flat-bottomed thick-walled aluminum

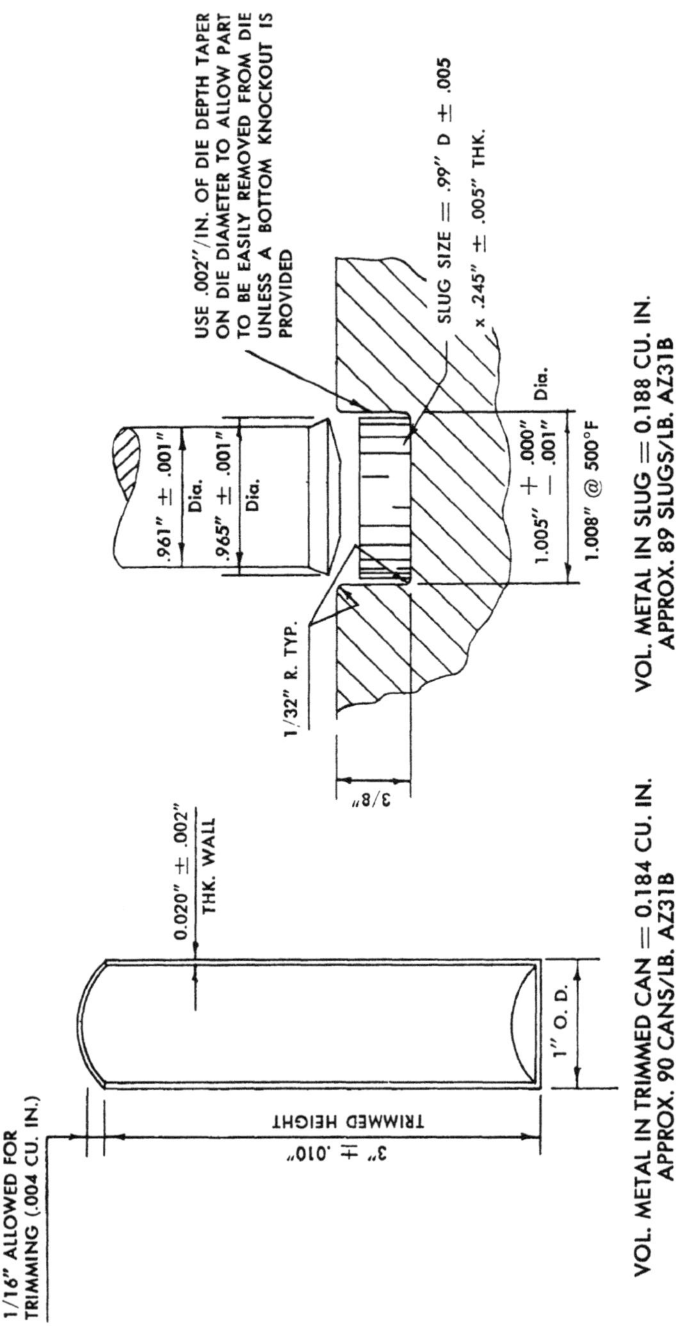

FIG. 4:3. TOOLS USED FOR BACKWARD EXTRUSION OF MAGNESIUM (*courtesy* Dow Metal Products Company)

shells. Figure 4:2*b* shows a similar punch-and-die combination used for the production of zinc shells. The major differences from the tools for aluminum lie in the use of a greater bevel angle and a longer chamfered end. Figure 4:3 shows tooling recommended for the production of magnesium shells. These

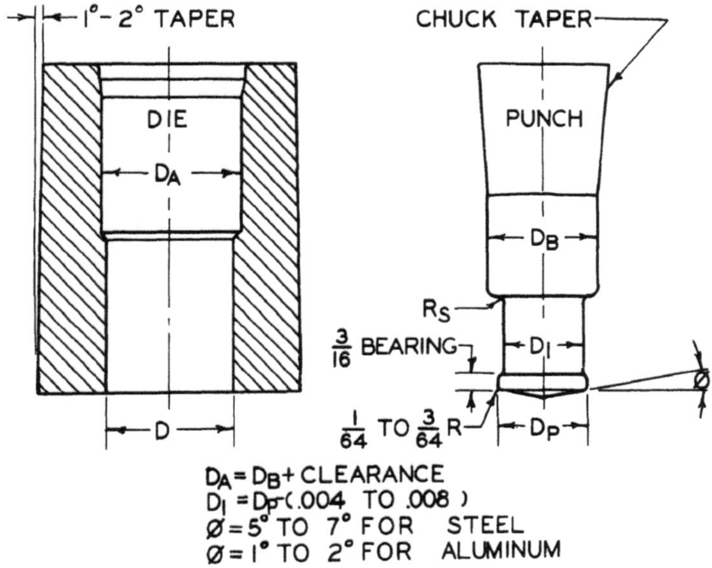

FIG. 4:4. TOOLS USED FOR BACKWARD EXTRUSION OF STEEL (*courtesy* Parker Rust Proof Division, Hooker Chemical Corporation)

tools are used with heated blanks. Figure 4:4 shows a set of tools recommended for the extrusion of steel parts, but applicable to aluminum with the modification indicated. A comparison of this design with that of Figure 4:3 indicates the increased sturdiness of the punch to compensate for the higher pressures involved.

Designing the Die

Whereas the punch is stressed in compression, the die is stressed in tension. These stresses are high but they vary with the strength of the alloy being extruded. Among the strongest materials extruded are the steels. Morgan[14] calculated that the tensile hoop stress on the inner face of an unsupported die reaches 120 tons to the square inch in the extrusion of steel with a 50 per cent reduction of area.

Because of the high stresses, the dies expand elastically during extrusion. As a backward-extruded part is produced with a moving orifice, and the region of maximum pressure moves downward in the die as the punch descends, the expanding region also moves down from the mouth. As the region of high pressure descends, the mouth of the die recovers and choking

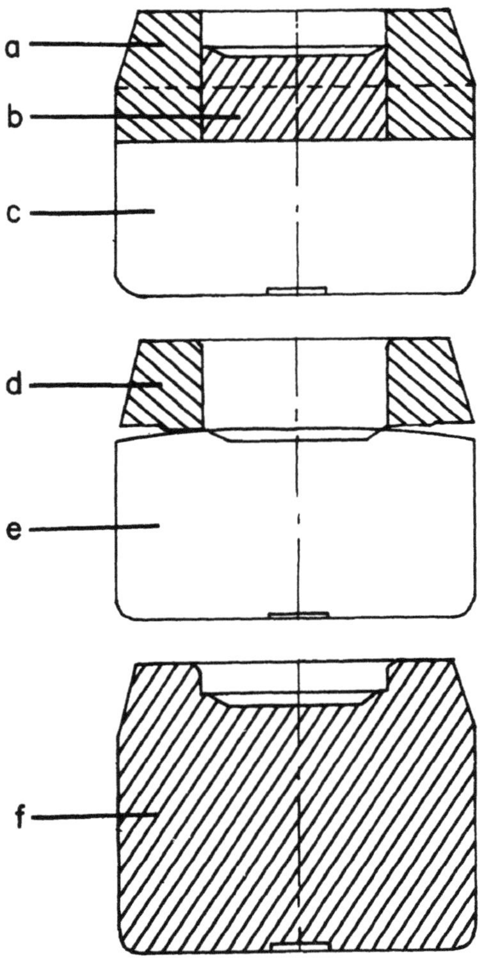

FIG. 4:5. DESIGNS OF DIES FOR BACKWARD EXTRUSION OF ALUMINUM
Top, Three-piece Die; *a*) Ring, *b*) Pad, *c*) Base
Middle, Two-piece Die; *d*) Ring, *e*) Base
Bottom, One-piece Die; *e*) Ring and Base combined
(Lang *et al*[6])

will result unless provision has been made in the design to compensate for this recovery.

This phenomenon is one of the fundamental properties of the backward-extrusion die. As a direct result, the walls are subjected to alternating stresses during the production of each part. The failure of dies during long production runs is frequently caused by fatigue.

Although the inside diameter of an extruded part is determined by the diameter of the extrusion edge on the punch, the thickness of the wall, and hence the outside diameter, depend on the expansion of the die. If the clearance between the extrusion edge and the die at rest is compared with the wall-thickness of a part, it will be found that the latter is greater than expected. This dimensional change must be considered if parts are to be produced within desired limits of tolerance.

Figure 4:5 shows typical designs for the dies used in the production of nonferrous parts by backward extrusion. The simplest die is the one-piece design shown at the bottom. This type of die is used frequently but it has certain disadvantages. It is not so easy to produce as the other designs and failure at any point means discarding the entire die. The dimensions, however, are similar to those discussed below for the two-piece die.

The two-piece die, shown in the center, is much more popular because one part of it can be renewed without discarding the other. Gabler[8] points out that failures occur in the base of a die more quickly than in the ring; the latter can be used three to four times as long as the former. Because of this fact, the three-piece die shown at the top is favored by some producers.

Recommendations from several sources have been followed to develop the dimensions of a typical two-piece die for the production of an aluminum part.[6,8,10] The base of the die may be flat but preferably it is slightly concave. Corner radii should be not less than 0.040 inch and the cavity should be as shallow as possible. The outside diameter of the base should be approximately $2\frac{1}{2}$ times the diameter of the can and the height should be equal to the diameter of the can.

The ring of the die should align accurately with the inside diameter of the cavity in the base and it should be slightly larger—but not over 0.002 inch. The inside of the ring should be parallel for a distance of about $\frac{1}{8}$ inch greater than the thickness of the blank. Above this point, the ring should be given a slight outward taper and finally opened on a radius of about two degrees at the mouth.

To ensure tightness in clamping, the outside of the bottom of the ring should be prepared with a clearance, as shown in Figure 4:5. A design of this type (*d*) will cause the entire clamping pressure to be applied on the inner part of the ring. A 20-degree taper on the outside of the ring assists in clamping.

The thickness of the wall of the extruded part is determined by the clearance between the extrusion edge and the face of the die, with a suitable allowance for expansion of the die. Galloway[10] states that in designing tools for the production of aluminum cans, the thickness of the wall should be no less than one per cent of the diameter of the can and the base should be somewhat thicker than the wall.

Because of the alternating stresses mentioned previously, all corners should be rounded and there should be no abrupt changes in section either in the punch or in the die. As suggested for the punch, all surfaces should be kept highly polished.

In discussing dies for extrusion of steel, Morgan[5] suggests that the walls of the die should be parallel for two-thirds of the depth of the cavity and should taper to a mouth that is about 0.004 inch larger than the diameter of the die. Howard *et al.*[3] state that solid dies are susceptible to cracking at the junction of the wall and base during the extrusion of steel. Examples of shapes of dies are shown in Figures 4:2 to 4:4.

To enable the die to withstand the high tensile stresses involved without bursting, it must be supported. This is accomplished by the use of compression rings. According to Galloway[13] the size of shrink rings should be selected to give the optimum distribution of stress across the various members. For a single ring, this distribution can be achieved by using the following formula:

$$D_d = D_a \times D_i R$$

where D_d = joint diameter
D_a = outside diameter
D_i = inside diameter
R = ratio of yield strengths of the two members

Galloway notes that this type of construction is particularly adaptable to dies made of cemented carbide, which show excellent resistance to wear and give good tool life when they are properly supported.

Sieber[15] also suggests using a thin-walled die with a shrunk-on outer ring to reduce the peak stress on the inner wall. It can be further reinforced by using a second ring. He states that the relative proportions of the die and the shrink rings are usually

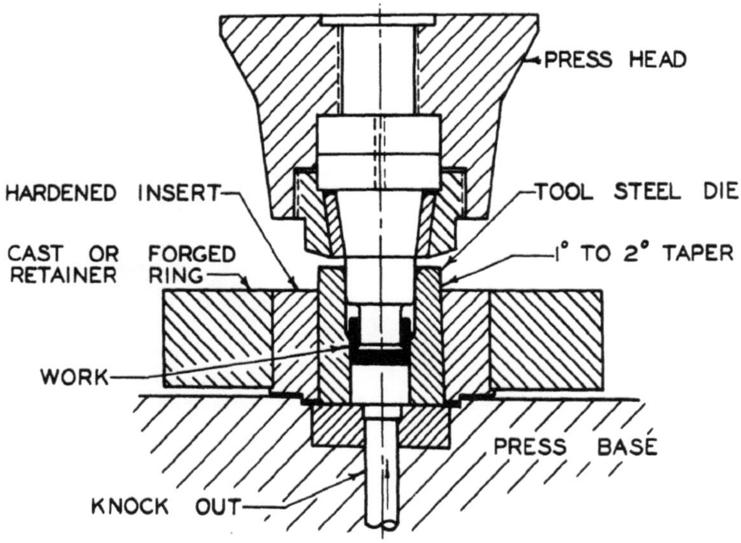

FIG. 4:6. TOOLING ASSEMBLY FOR BACKWARD EXTRUSION (*courtesy* Parker Rust Proof Division, of Hooker Chemical Corporation)

calculated by Lamé's formula for the stresses in thick-walled cylinders. Sieber notes that the relative strengths of unsupported dies and of dies singly and doubly reinforced are in the proportion 1 to 1.45 to 1.80. For the extrusion of steel, shrink rings are designed to develop a compressive hoop stress of about 70 tons per square inch on the die.

Typical tooling assemblies for backward extrusion are shown in Figures 4:6 and 4:7; the latter is used for the extrusion of heated blanks.

MODIFIED BACKWARD EXTRUSION

In modified backward extrusions the punch is fixed in position and the base of the die acts as the moving member of the tool combination. The method has the advantage of minimizing floating of the punch. Land *et al.*[6] reported that this modification had been used in Germany during World War II but they gave merely a sketch of the tooling with no details of

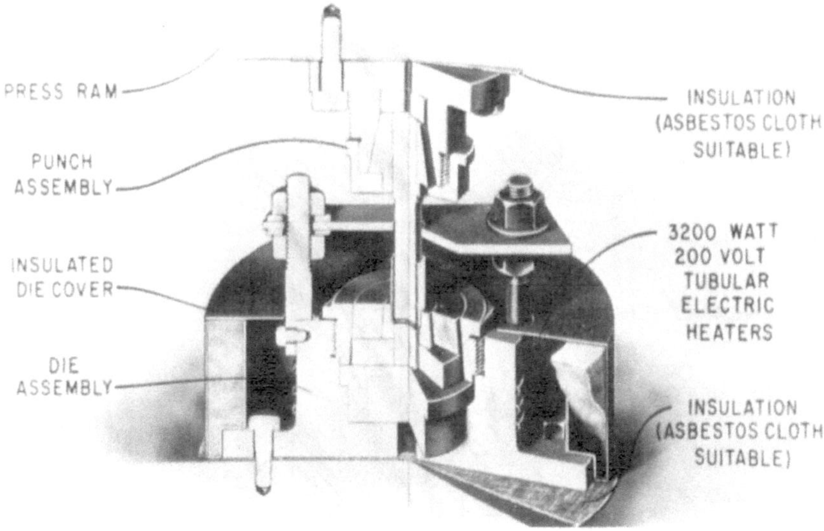

FIG. 4:7. TOOLING ASSEMBLY FOR BACKWARD EXTRUSION OF MAGNESIUM (*courtesy* Dow Metal Products Co.)

its use. The method has been employed by Kunogi[16] for the extrusion of steel and he stated that the extrusion pressure was reduced greatly by this procedure. A sketch of his tooling arrangement is given in Figure 4:8.

FORWARD EXTRUSION

Whereas in backward extrusion the metal flows in a direction opposite to that of the punch, in forward extrusion the flow is in the same direction. Length of punch is virtually independent of length of part and a short sturdy punch can be used. This permits an accurate alignment of punch and die at all times and reduces problems of operation.

Another difference between the two procedures is length of part. In backward extrusion, the length is limited by the buckling strength of the punch. In forward extrusion, the only limitations are the size of the blank required and the clearance necessary to remove the part from the die. When the stroke of the punch is completed a butt-end remains in the die. The part can be removed either by forcing it back through the die after removing the punch, or by inserting another blank and extruding it.

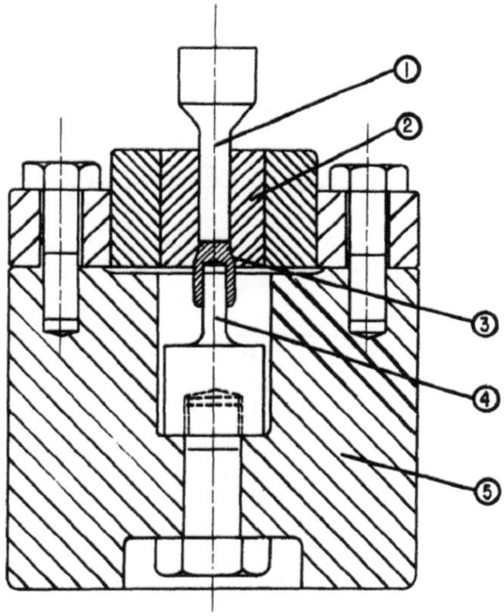

FIG. 4:8. TOOLS FOR MODIFIED BACKWARD EXTRUSION
1) Plunger; 2) Die Ring; 3) Extruded Part; 4) Punch; 5) Base
(*after* Kunogi[16])

Design of Punch

The function of the punch in forward extrusion is to compress the material and force it to flow through a restricted opening in the die. Only in hollow extrusions does it contribute to the shape of the part.

The punch used for producing a solid part is generally quite simple, its major requirement being a close fit in the die to prevent formation of flash. The clearance should be about 0.004 to 0.008 inch.

For the extrusion of a tubular part, the punch is somewhat more complicated although its design is not so critical as it is for the one used for backward extrusion. The punch is designed to permit an extension to enter the cavity of the die and act as a mandrel to establish both the thickness of the wall and the contour of the part. The extension must be designed to come into contact with the bottom of the blank at the instant that the punch starts to compress the material. If the extension strikes the bottom first, the tools may be damaged. If extrusion starts first, the shape may not be maintained. To prevent collapse of closed-end parts, it is often necessary to bleed air through the punch into the cavity that forms as extrusion progresses. To reduce friction, a taper of a few thousandths of an inch should be provided on the extension of the punch.

A set of tools that has been recommended for the forward extrusion of steel is shown in Figure 4:9. Dimensions are indicated on this figure.

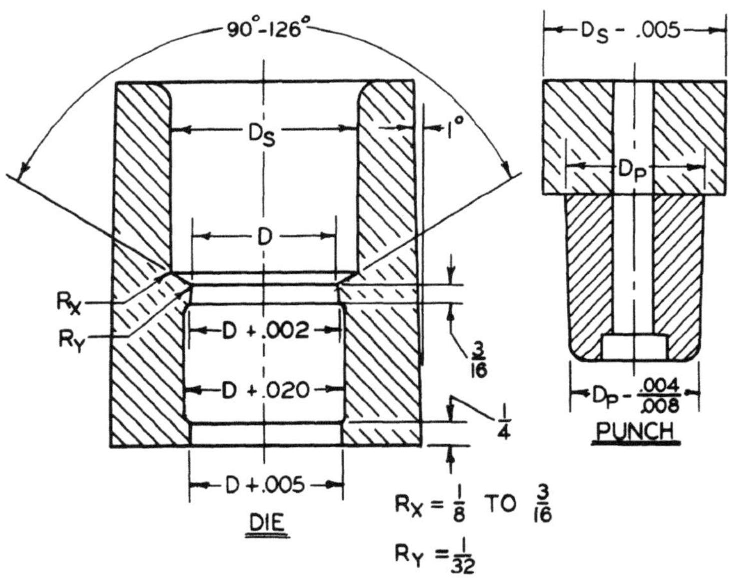

FIG. 4:9. TOOLS FOR FORWARD EXTRUSION
(*courtesy* Parker Rust Proof Division, Hooker Chemical Corporation)

Design of Die

The design of the die for forward extrusion is more critical than that of the punch. Generally, a one-piece construction is used, as shown in Figure 4:9. The mouth is rounded slightly to guide the punch. The cavity in the die must be long enough to hold the blank and permit entry of the punch before extrusion starts. The walls are parallel and terminate at a shoulder having the desired extrusion angle. The junction between side walls and shoulder should be rounded but there should be a relatively sharp angle at the junction of the shoulder and the extrusion edge. The extrusion edge should be short and the die block, below the edge, should be somewhat enlarged to reduce friction. A design suggested for the extrusion of steel is shown in Figure 4:9.

The most important features of dies for forward extrusion are the angles of the shoulders and the radii leading into the extrusion edge. For large reductions of area, a flat included angle

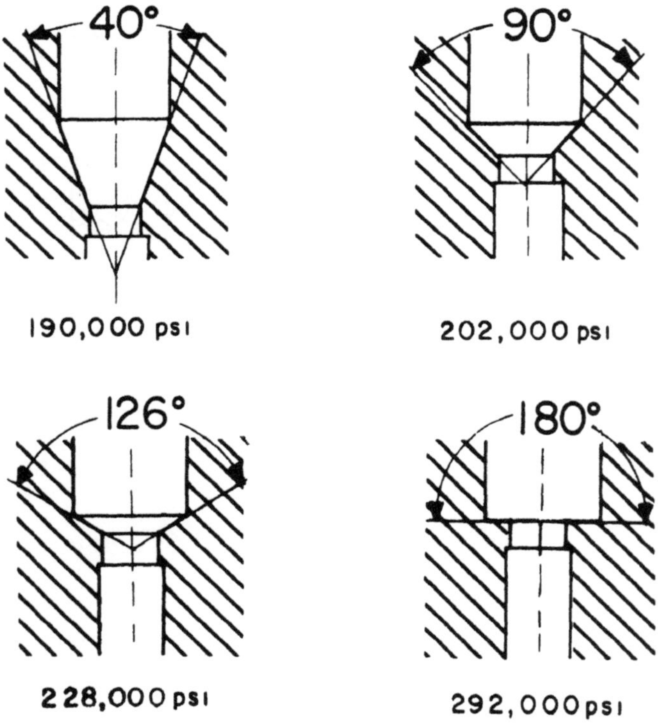

Fig. 4:10. Effect of Angle of Die on Pressure During Forward Extrusion of Steel (Hauttmann and Pessl[17])

and a sharp radius combine to form a clearance between part and die with the result that friction is reduced and the lubricant is retained. Smaller included angles and larger radii can lead to galling of the part during extrusion and sticking during ejection from the die.

The effect of the shoulder angle on the extrusion pressure is indicated in Figure 4:10 based on the work of Hauttmann and

Pessl.[17] The pressures given are averages of tests on eleven steels ranging from 0.08 to 0.50% carbon, and with tensile strengths ranging from 50,000 to 110,000 psi. They determined that the lowest extrusion pressure was required for an orifice with a small angle. Such an angle also resulted in the most favorable conditions for flow. The radial stresses are greatest in such designs, however, and they increase the danger of failure of the die. With a reduction in cross-sectional area of

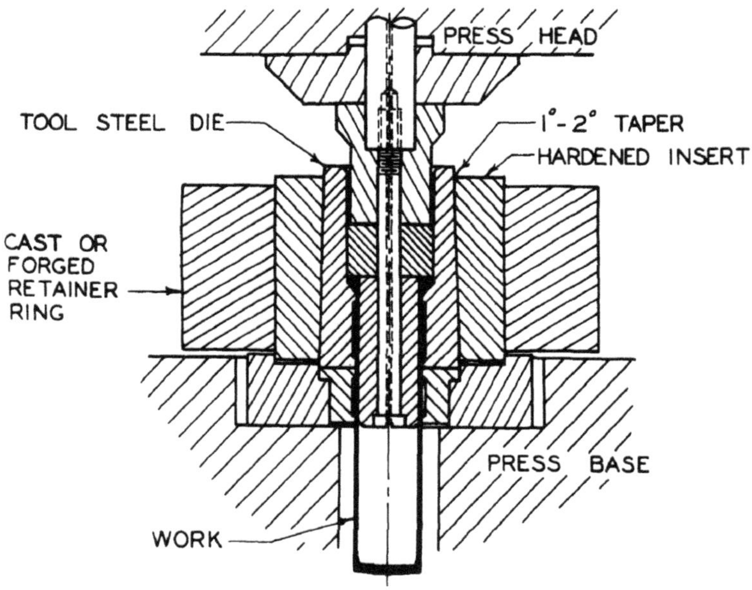

FIG. 4:11. TOOLING ASSEMBLY FOR FORWARD EXTRUSION (*courtesy* Parker Rust Proof Division, Hooker Chemical Corporation)

65 per cent, the extrusion pressure averaged 2.4 times the tensile strength when the angle of the die was 40 degrees; 2.7 for 90 degrees; 3.1 for 126 degrees; and 3.9 for 180 degrees. They concluded that to minimize stresses on the die the most favorable shoulder angles for steel range from 125 to 130 degrees.

According to Boston,[9] the diameter of a die for forward extrusion is based on the area of the blank. For extrusion of aluminum, he suggests that the diameter of the die be $2\frac{1}{2}$ inches for a blank area of one square inch; 12 inches for an area of 7 square inches. The minimum height of the block for the die should be $2\frac{1}{2}$ times the diameter of the blank and the depth of the cavity should equal the thickness of the blank plus $\frac{3}{16}$ inch. Most of the recommendations for dies for extrusion of steel suggest using a relatively thin die block backed with shrink rings. A typical tooling assembly for forward extrusion is illustrated in Figure 4:11.

COMBINED FORWARD AND BACKWARD EXTRUSION

By a suitable combination of the tools used for backward and forward extrusion, parts can be extruded in both directions simultaneously. For example, in producing a collapsible tube the neck is formed by forward extrusion and the side-walls by backward extrusion. Typical tools for the production of such a part are shown in Figure 4:12. Comparison of these tools with those shown in Figures 4:2 and 4:9 indicates that certain features of each type of extrusion tool have been incorporated. Thus, the extrusion edge on the punch is a feature of backward extrusion, whereas the tip on the punch is a feature of the forward extrusion of a hollow component.

SELECTION OF PRESS

Selection of the type of press to be used for impact extrusion is a matter of considerable controversy. Some producers prefer mechanical presses, others select hydraulic. In a questionnaire to producers of parts from various nonferrous alloys and steels, twelve answers were received on a preference in presses. Nine favored mechanical presses, giving as their reasons such factors as lower capital costs, lower operating costs, easier maintenance, and higher production rates. Two preferred hydraulic presses, citing as the reasons for their choice better flow of metal in relation to speed, and greater flexibility. One producer had no preference; in his opinion, parts of equal quality could be produced on either type of press. There was no relation between the alloys being extruded and the type of press selected.

It would appear that, in many cases, a company starting to produce parts by impact extrusion has presses available that have been used for other forming operations. The logical procedure is to adapt these presses to the job, at least in the early stages of production. As a result, practically all types of mechanical presses have been used for the production of parts from soft metals and aluminum alloys. For example, Boston[9] cites the use of standard inclinable and straight-side punch presses for the extrusion of aluminum parts.

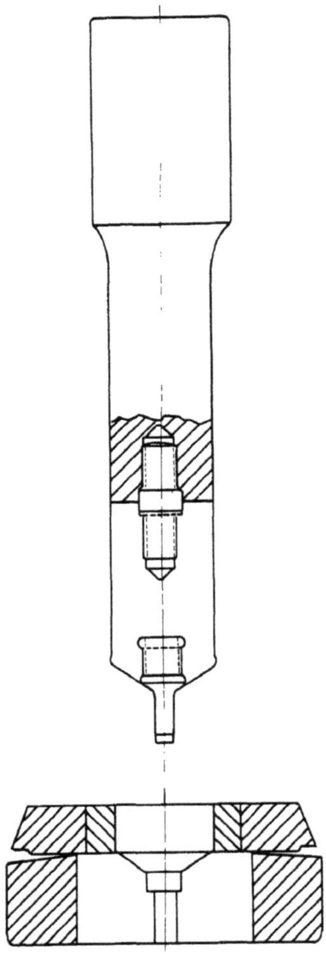

FIG. 4:12. TOOLS USED IN GERMANY FOR PRODUCTION OF COLLAPSIBLE TUBES (Lang *et al*[6])

Although greater pressures are required for the extrusion of steel parts than for the corresponding aluminum parts, the former have been produced successfully on general-purpose machines of both the mechanical and the hydraulic types.

The differences that characterize mechanical and hydraulic presses, in the methods of applying the load, do not appear so significant as to prevent the use of either type of press. Apparently, the major requirements are rigidity and sufficient power to extrude the part desired from the specified material.

In recognition of the fact that extreme rigidity is a desirable feature, new presses have been designed specifically for impact or cold extrusion. Both mechanical and hydraulic types are available from various producers.

Selection of the press has been discussed in a number of articles in various publications and they might be consulted for details of the features of each type and their advantages and disadvantages. Two excellent discussions of the subject are those of Johansen[18] and Massey.[19]

REFERENCES

1. GARDNER, R. W., Techniques and Material for Cold Extrusion, Paper No. 36A, SAE National Production Meeting, March 1958.
2. HOWARD, F., Designing with Impact Extruded Components, *Engineering Materials and Design*, October 1958:26.
3. HOWARD, F., DENNISON, H. A. J., and ANGUS, N., Some Investigations Into the Cold Extrusion of Steel, *Sheet Metal Industries*, 38:403 (1961).
4. VON ZEERLEDER, A., and VON BERG, E., Relationship of the Force Requirements in the Extrusion of Tubes of Pure Aluminum and Various Aluminum Alloys, *Schweizer Archiv*, 3:109 (1937).
5. MORGAN, R. A. P., The Cold Extrusion of Steel, *J. Iron Steel Inst.* (London), 193:285 (1959).
6. LANG, R. F., ANSELL, W. R., and WILSON, D. V., Impact Extrusion: German Practice 1946, *B.I.O.S. Final Report* No. 1405, (1946).
7. FISCHER, H., The Cold Extrusion of Steel, *Sheet Metal Industries*, 30:447 (1953).
8. GABLER, P., Precision Impact Extrusion and Die Casting of Aluminum in Industry, *Aluminum*, 25:194 (1943).
9. BOSTON J. R., Cold Extrusion of Aluminum Parts for Douglas Aircraft, *Machinery*, 51, July 1945:138.
10. GALLOWAY, K. A., The Impact Extrusion of Aluminum, *Light Metals*, April 1953:125.
11. Private Communication, Dow Metal Products Company, July 1962.
12. LELAND, J. F., Tooling for Cold Steel Extrusion, *Machinery* July 1955:170.
13. GALLOWAY, D. F., Cold Extrusion—A Review of Some Technical Information and Current Practice, *Microtechnic*, 15:85 (1961).
14. MORGAN, R. A. P., Discussion on Cold Extrusion, *J. Iron Steel Inst.* 195:324 (1960).
15. SIEBER, K., Developments in the Cold Flow Pressing and Extrusion of Steel, *Machinery* (London), 81:937 (1952).
16. KUNOGI, M., New Method of Cold Extrusion of Steel, *Trans. Japan Soc. Mech. Engineers*, 23, October 1957:742.
17. HAUTTMANN, H. H. and PESSL, H. J., The Aluminum-Bearing Non-Aging Steels and Their Cold Extrusion, *Steel Processing*, November 1950:557.
18. JOHANSEN, E. K., Mechanical Press Equipment for the Cold Extrusion of Steel, *Sheet Metal Industries*, June 1953:476.
19. MASSEY, T. F., The Relative Merits of Presses for the Cold Extrusion of Steel, *Ibid.*, 479.

Chapter 5

TOOL MATERIALS

In selecting a steel for tools, the most important factor is generally the mechanical requirements of the tool. Other factors that may be considered are availability of the steel, the design of the tool, length of the production run, and cost of the grade of the steel. The aim is to select a material that will assure the lowest unit cost of production.

The principal characteristics of impact extrusion, as far as tooling is concerned, are the high compressive loads on the punch and the high tensile loads on the die. Punches must have sufficient compressive strength to resist upsetting without being dangerously brittle. Dies require high tensile yield strength even though they are invariably prestressed in compression by means of shrink rings. These requirements of the punch and die are such that it is necessary to use tool steels that will harden throughout in the required size of the section.

TYPES OF TOOL STEELS

Tool steels are marketed under more than 1200 trade names. However, the AISI and SAE[1] have classified them into six groups:

- Water-Hardening
- Shock-Resisting
- Cold Work
- Hot Work
- High Speed
- Special Purpose

Each group is identified by a letter and may be subdivided into a number of individual types, each identified by a number following the letter. The composition indicated for each type is intended for identification only. Steels of the same type, produced by different steel companies may vary in the content of the listed elements and may contain elements not listed in the identifying composition.

Although it is probable that steels from most of these groups have been used at one time or another for the production of tools for impact extrusion, certain steels are especially recommended for the purpose. Table 5:1 lists the types that are suggested by the major producers of steel for tools. Applications of specific types, based on the suggestions of producers, are given in Table 5:2. The groups are distinguished as follows:

Water-Hardening Tool Steels (Symbol W)

Although the water-hardening steels are nominally carbon steels, they may contain small additions of chromium or vanadium to increase the depth to which the steel will harden. They are shallow-hardening. In sections over one-half inch, quenching in water produces a hard case and a tough core.

Shock Resisting Tool Steels (Symbol S)

The shock-resisting steels usually contain moderate additions of silicon, chromium, and tungsten, and sometimes molybdenum or nickel. Carbon is maintained around 0.50 per cent to give the steels high strength with moderate resistance to wear. Type S1 is relatively deep-hardening in oil and is relatively ductile when hardened to Rockwell C60.

Cold Work Tool Steels: Oil-Hardening Type (Symbol O)

The oil-hardening type of cold work steels contain manganese, chromium, and tungsten; sometimes molybdenum is added. Their high carbon content produces sufficient wear-resistance for short-run applications. Type O1 can be hardened to Rockwell C60 to the center of a $1\frac{1}{4}$-inch round. Whereas Type W1 will harden to Rockwell C55 only to a depth of about one-fourth inch, Type O1 will harden to the same degree to the center of a two-inch round.

Cold Work Tool Steels: Medium Alloy, Air-Hardening Type (Symbol A)

The air-hardening cold work steels usually contain sufficient manganese, chromium, molybdenum, and vanadium to promote deep hardening. For example, Type A2 can be hardened to Rockwell C62 to the center of a 3½-inch round. The specified high carbon content contributes resistance to wear. These steels are less expensive than the Type D steels but have approximately the same resistance to distortion during the heat treatment. Like the Type D steels, they resist softening at elevated temperatures.

TABLE 5:1
TOOL STEELS SUGGESTED FOR IMPACT EXTRUSION*

AISI Type	*Identifying Composition, %*						
	C	Mn	Cr	Mo	W	V	Co
			Water-Hardening Tool Steels				
W1	0.60-1.40	—	—	—	—	—	—
W2	0.60-1.40	—	—	—	—	0.25	—
W3	1.00	—	—	—	—	0.50	—
W4	0.60-1.40	—	0.25	—	—	—	—
			Shock-Resisting Tool Steels				
S1	0.50	—	1.50	—	2.50	—	—
			Cold Work Tool Steels: Oil-Hardening				
O1	0.90	1.00	0.50	—	0.50	—	—
			Cold Work Tool Steels: Medium Alloy, Air-Hardening				
A2	1.00	—	5.00	1.00	—	—	—
A4	1.00	2.00	1.00	1.00	—	—	—
			Cold Work Tool Steels: High Carbon, High Chromium				
D1	1.00	—	12.00	1.00	—	—	—
D2	1.50	—	12.00	1.00	—	—	—
D3	2.25	—	12.00	—	—	—	—
D5	1.50	—	12.00	1.00	—	—	3.00
			Special Purpose Tool Steels				
F2	1.25	—	—	—	3.50	—	—
F3	1.25	—	0.75	—	3.50	—	—
			High Speed Tool Steels: Molybdenum Base				
M2	0.80	—	4.00	5.00	6.00	2.00	—
M3	1.00	—	4.00	5.00	6.00	2.70	—
			High Speed Tool Steels: Tungsten Base				
T1	0.70	—	4.00	—	18.00	1.00	—

* Based on recommendations in producers' catalogs

Cold Work Tool Steels: High-Carbon, High-Chromium Type (Symbol D)

The principal alloying element in the series of high-carbon chromium steels is chromium but they may also contain molybdenum, tungsten, cobalt, and vanadium. These steels are deep-hardening and undergo minimum distortion during heat-treatment. As Type D2, for example, hardens to a depth of at least ten inches, hardenability is not a significant factor in the use of these steels in tools for impact extrusion. Steels of this group are highly resistant to wear and to softening at elevated temperatures.

TABLE 5:2

APPLICATIONS OF PRINCIPAL TYPES OF TOOL STEELS

Objective	*For Dies*	*For Punches*
Short-run, Minimum Initial Cost	W1, W2, F2, O1	F2, W2, W4
General Purpose	D5, D3, A2, F3	O1, A2, A4
Long-run, High Production	D5, T1, M2, M3	D2, D5, S1, M2, M3

Special Purpose Tool Steels (Symbol F)

The principal alloying element in special-purpose steels is tungsten, although chromium may be added also. These are shallow-hardening steels requiring quenching in water unless chromium is added. They are generally more resistant to wear than the Type W steels. Type F2, for example, shows a lower hardenability than W steels of the same carbon content but the hardness of the core is higher; in a one-inch round the core of an F steel would harden to Rockwell C45-50, whereas that of a W steel would harden only to Rockwell C40-45.

High-Speed Tool Steels: Molybdenum Base (Symbol M)

The principal alloying elements in the molybdenum steels for high-speed tools are molybdenum, tungsten, chromium, vanadium, and cobalt. They are deep-hardening steels having properties similar to those of the tungsten high-speed steels but they are tougher for the same hardness. They are considerably less expensive than the tungsten grades and are displacing the latter in many applications.

High-Speed Tool Steels: Tungsten Base (Symbol T)

The principal alloying elements in the tungsten steels on high-speed tools are tungsten, chromium, vanadium, and cobalt.

TABLE 5:3

RECOMMENDED HEAT TREATING PROCEDURES FOR STEELS LISTED IN TABLE 5:1

AISI Type	*Annealing*		*Hardening Conditions*					*Tempering*
	Temp., F	Cooling Rate, F/hr	Heating Rate	Preheat Temp., F	Hardening		Quenching Medium[a]	Temp., F
					Temp., F	Time, min		
W1	1360-1450	75	Slow	—	1450-1500	10-30	B or W	300-650
W2	1360-1450	75	Slow	1050-1200[b]	1400-1550	10-30	B or W	300-650
W3	1360-1450	75	Slow	1050-1200[b]	1400-1550	10-30	B or W	300-650
W4	1360-1450	75	Slow	1050-1200[b]	1400-1550	10-30	B or W	300-650
S1	1450-1500	50	Slow	—	1650-1800	15-45	O	400-1200
O1	1400-1450	50	Slow	1200	1450-1500	10-30	O	300-500
A2	1550-1600	40	Slow	1450	1700-1800	20-45	A	350-1000
A4	1360-1400	25	Slow	1250	1500-1600	15-90	A	300-800
D1	1600-1650	50	Very slow	1500	1775-1850	15-45	A	400-1000
D2	1600-1650	50	Very slow	1500	1800-1875	15-45	A	400-1000
D3	1600-1650	50	Very slow	1500	1700-1800	15-45	O	400-1000
D5	1600-1650	50	Very slow	1500	1800-1875	15-45	A	400-1000
F2	1450-1500	50	Slow	1200	1450-1600	15	B or W	300-500
F3	1450-1500	50	Slow	1200	1450-1600	15	W,B,O	300-500
M2	1600-1650	50	[c]	1350-1550	2175-2250	2-5	O,A,S	1000-1100
M3	1600-1650	50	[c]	1350-1550	2200-2250	2-5	O,A,S	1000-1100
T1	1600-1650	50	[c]	1500-1600	2300-2375	2-5	O,A,S	1000-1100

[a] *W* = water; B = brine; O = oil; A = air; S = salt bath
[b] For large tools
[c] Rapid after preheating
NOTE: Double tempering is suggested for Types M2, M3, and T1

They are deep-hardening steels which can be hardened to at least Rockwell C64 in oil or in hot salt. They have high resistance to wear but are not so tough as the molybdenum high-speed steels.

HEAT TREATMENT

As mentioned before, the tool steels fall generally into three classes: 1) water-hardening, 2) oil-hardening, and 3) air-hardening. The method of hardening has certain effects on the product obtained as discussed briefly in the following paragraphs:

Water-Hardening Steels

Steels that harden in water are shallow-hardening with the result that, excepting in very thin sections, heat treatment produces a hard, wear-resistant case with a softer shock-resistant core. As the water quenching required introduces high internal stresses, these steels are not adaptable to tools having sharp corners or abrupt changes in section because of the danger of cracking during heat treatment. Considerable distortion may occur also during heat treatment.

Oil-Hardening Steels

The oil-hardening steels contain sufficient alloying elements to be through-hardening in relatively heavy sections, and therefore have no shock-resisting core. They are less sensitive to sharp corners and abrupt changes in section than the water-hardening steels and less distortion occurs during heat treatment.

Air-Hardening Steels

The air-hardening steels deform less than the oil-hardening steels during heat treatment; although they are frequently called nondeforming steels, there is some dimensional change. The air-hardening steels are more resistant to wear than the oil-hardening grades. They require higher hardening temperatures and longer soaking times, however, and the surface must be protected during heating for hardening.

Heat-treating procedures recommended for the steels listed in Table 5:1 are given in Table 5:3. These suggestions indicate the ranges of temperature and the procedures required to attain the properties desired in these steels. All of them, except the high-speed steels, should be heated slowly to the hardening

TABLE 5:4

EFFECT OF TEMPERING ON THE HARDNESS OF SOME TOOL STEELS

AISI Type	*Hardening,* F	*Rockwell C Hardness after Tempering at* 300 F	400 F	500 F	600 F	800 F	1000 F	1200 F
W1[a]	1475, WQ	64-66	62-64	58-60	54-56	46-48	35-39	—
W1[b]	1550, Brine	64-66	60-62	57-59	52-54	42-44	—	—
O1	1475, OQ	63-65	60-63	58-60	55-57	48-50	40-44	—
A2	1775, AQ	63-64	60-62	58-60	56-59	56-59	56-58	—
A4	1550, AQ	60-62	58-60	57-59	56-58	52-54	47-48	—
D2	1850, AQ	61-63	59-61	57-59	56-58	57-59	56-58	43-45
D3	1800, AQ	63-65	62-64	61-63	59-61	56-58	52-54	43-44
D3	1800, OQ	62-63	61-62	59-60	58-59	56-57	—	—
F2	1575, WQ	64-66	63-65	61-63	58-60	51-53	44-46	35-37
F3	1600, Brine	64-66	63-65	61-63	59-61	—	—	—
M2	2250, OQ	64-66	63-66	62-64	62-63	63-64	65-66	53-54
T1	2350, OQ	63-64	63-64	62-63	61-62	61-62	63-64	56-59

[a] 1.05% carbon

[b] 0.95% carbon

NOTE: WQ = water-quenched; OQ = oil-quenched; AQ = air-quenched temperature, held there for sufficient time to equalize the temperature throughout the tool, and then quenched. As indicated in the table, some steels should be heated at an intermediate temperature before heating to the hardening temperature.

The rate of heating for hardening should be slower for alloy tool steels than for carbon tool steels. In general, the slower and more uniform the heating, the less change in size or shape occurs. The high-speed steels must be heated slowly to the temperature for preheating, held for sufficient time to equalize the temperature, then heated rapidly to the quenching temperature, and quenched after holding this for a short time. A protective atmosphere is desirable for all these steels; and essential for those steels that are heated to the higher temperatures. It should be noted that Table 5:3 is a generalized table; in using a specific steel, the heat treatment recommended by the producer should be followed.

After quenching, the steels should be tempered immediately at the temperature required to attain the desired hardness. Table 5:4 indicates the Rockwell C hardness values of representative steels resulting from tempering at the indicated temperatures. The values given in this table are not based on individual steels but are ranges obtained from various published sources; therefore, they are merely indicative of the ranges to be expected.

TOOLING FOR SPECIFIC APPLICATIONS

The selection of material for a tool depends to a considerable degree on the characteristics of the metal being extruded. Lead and tin are low-strength materials which make the least demand on the tools. Tube grade aluminum is next in strength followed by commercial aluminum and the aluminum alloys. Demands on the tools are considerably more severe for the production of impact extruded items from these alloys, particularly from the high-strength alloys, such as 7075. The steels probably make the greatest demands on tool materials.

Lead and Tin

Little information has been published on the requirements for tooling in the impact extrusion of items from lead and tin. In a survey made by sending a questionnaire to a number of producers of lead and tin components, a scattering of replies indicated that dies were being produced from Types W1 and F2 steels, punch shafts from Types O1 and A4, and punch tips from Types D1 and S1. The purpose of making the punch tip from Type D1 was to resist mushrooming of the tip.

Lang, Ansell, and Wilson[2] reported the analyses given in Table 5:5 for a typical set of extrusion tools used for soft metals in Germany. Both punch and die were made of high-carbon-chromium-vanadium steels, although the content of alloy in the punch was somewhat higher than that in the die. The punch was hardened to Rockwell C60, the die to Rockwell C63. The pad was made of a high-carbon steel and hardened to Rockwell C48.

Aluminum and Aluminum Alloys

A survey of producers of impact-extruded items from aluminum indicated that 12% chromium steels were favored by a number of them for both punch and die, although several mentioned the use of tungsten carbide inserts for the die.

These answers confirm the recommendations of several others who have been somewhat more specific in their comments. For example, Gabler[3] states that 12%-chromium steels should be used for both punch and die. Punches hardened to Rockwell C58-60 should produce 200,000 parts from commercial aluminum. Dies hardened to Rockwell C60-62 should produce 50,000 to 70,000 parts from the same material.

For the extrusion of commercial aluminum and an aluminum alloy containing one per cent manganese, Galloway[4] used a 2%-carbon, 14%-chromium steel hardened to Rockwell C60 for both punch and die ring. The die base (or pad) was made from a chromium-tungsten steel hardened to Rockwell C54 and the shrink ring, from a $4\frac{1}{4}$% -nickel steel hardened to Rockwell C47.

Boston[5] suggested use of a 2%-carbon, 13%-chromium steel for both punch and die in the production of parts from aluminum alloys, the punch being hardened to Rockwell C58-60, and the die, to Rockwell C55-57.

Shoemaker[6] recommended a $1\frac{1}{2}$%-carbon, 12%-chromium, 1%-molybdenum steel, hardened to Rockwell C58-60, for the extrusion of commercial aluminum and the annealed alloys 6061-0, 2014-0, and 7075-0. For the die to be used with these materials, he suggested a high-carbon-tungsten-molybdenum steel, hardened to Rockwell C56-58.

Lang, Ansell, and Wilson[2] reported that a number of German producers were using high-carbon, 4-6%-chromium steels during the war for the extrusion of commercial aluminum. In the extrusion of a milk can, they obtained 50,000 parts on a punch and 100,000 parts on a die ring. One company noticed a significant reduction in the life of tools during the extrusion of 99%- instead of 99.5%-aluminum. The Germans reported that improved tool life resulted from use of a high-carbon, $8\frac{1}{2}$%-chromium steel.

Copper and Copper Alloys

Little information has been published on the impact extrusion of copper and copper alloys. Galloway[4] used a high-carbon, high chromium steel, hardened to Rockwell C60, for both punch and die for the extrusion of electrolytic tough-pitch copper at room- and at higher-temperatures. As shown in Table 5:5, the die base was made from a chromium-tungsten steel and the shrink ring from a nickel steel.

Magnesium and Magnesium Alloys

The impact extrusion of magnesium and its alloys differs from that of the other common nonferrous alloys because the magnesium blanks are invariably heated before extrusion, whereas those of the other metals are usually fed cold to the press. Normally, blanks of magnesium alloy are heated to about 500 F before extrusion.

Patton[7] states that the chromium-molybdenum-tungsten steel listed in Table 5:5 is suitable for the extrusion of commercial magnesium alloys. The steel is reputed to have good resistance to softening at elevated temperatures, a desirable feature since the temperature of the tools would rise above 500 F during commercial operation.

He noted that punches, hardened to Rockwell C55-60, have produced 200,000 parts before being discarded, and were usually discarded only because wear had reached the point at which the parts no longer remained within the required limits of tolerance. Dies were made either from the same steel as the punch and hardened to the same degree or were made from tungsten carbide. The latter material was preferred because the surface of the parts had better finish. Dies of tool steel had produced

TABLE 5:5
TOOL STEELS SUGGESTED FOR EXTRUSION OF VARIOUS NONFERROUS METALS

Part	*Composition, %*								*Rockwell C Hardness*	*Source (Ref)*
	C	Mn	Si	Cr	V	W	Mo	Other		
				Lead and Tin Alloys						
Punch	1.98	0.40	0.31	6.5	0.48	—	—	—	60[a]	2
Die Ring	1.99	0.49	0.29	5.2	0.26	—	—	—	63[a]	
Pad	1.22	0.28	0.22	—	—	—	—	—	48[a]	
				Aluminum and Aluminum Alloys						
Punch[b]	2	—	—	14	—	—	—	—	60	4
Die Ring[b]	2	—	—	14	—	—	—	—	60	
Die Base[b]	—	—	—	1.4	—	2	—	—	54	
Shrink Ring[b]	0.3	0.5	—	1.25	—	—	0.03	Ni 4.25	47[c]	
Punch	—	—	—	12	—	—	—	—	58-60	3
Die	—	—	—	12	—	—	—	—	60-62	
Punch	2	—	—	13	—	—	—	—	58-60	5
Die	2	—	—	13	—	—	—	—	55-57	
Punch	1.5	—	—	12	—	—	1	—	58-60	6
Die	1.5	0.5	—	—	—	2.8	0.5	—	56-58	
Punch	2.2	0.3	—	8.5	—	—	—	—	—	2
Die	2.2	0.3	—	8.5	—	—	—	—	—	
				Electrolytic Tough-Pitch Copper						
Punch[b]	2	—	—	14	—	—	—	—	60	4
Die Ring[b]	2	—	—	14	—	—	—	—	60	
Die Base[b]	—	—	—	1.4	—	2	—	—	54	
Shrink Ring[b]	0.3	—	—	1.25	—	—	0.03	Ni 4.25	47[c]	
				Magnesium and Magnesium Alloys						
Punch	0.35	—	1.0	5.0	—	1.25	1.5	—	55-60	7
Die	0.35	—	1.0	5.0	—	1.25	1.5	—	55-60	
				Commercial Titanium						
Punch[b]	0.7	—	—	4.0	1.0	18	—	—	62-64	8
Die[b]	1.5	—	—	12.0	1.0	—	1.0	—	59-60	

[a] Converted from Vickers hardness numbers
[b] Used for experimental work
[c] Estimated from tensile strength

about 200,000 pieces, whereas carbide dies had produced up to 10,000,000 pieces.

Titanium

The problems encountered in the impact extrusion of titanium appear to resemble those of steel more closely than those of the other nonferrous metals. The only tool materials reported for the extrusion of titanium were those used experimentally by Sabroff and his associates.[8] The steels and their hardness values are included in Table 5:5.

Carbon and Alloy Steels

Demands on tools for the cold extrusion of steel are much more severe than they are for most nonferrous metals. Punch loads run up to several hundred thousand *psi* and the obtaining of suitable tool materials was a major problem during the early years of development.

According to Howard[9] most punches and dies are produced from high-grade tool steels such as the high-carbon, high chromium types to withstand the stresses involved. Johnson and Bishop[10] report that the compressive strength of a 1.6%-carbon, 13%-chromium steel, oil-quenched from 1850 F and tempered at 300 F, was 425,000 *psi*. The compressive strength of other high-speed steels, of unspecified composition, was as high as 520,000 *psi*. By contrast, the compressive strength of a 1%-carbon tool steel, water-quenched from 1525 F and tempered at 300 F, was only 210,000 *psi* in the core, although in the case it was more than 400,000 *psi*. A punch made of this steel would fail very quickly.

Howard[9] mentioned also that conventional hardening and tempering to Rockwell C60-62, or even to C66, was advocated frequently but his company had obtained superior results by subzero treatments that involved cooling to −112 F for two hours after being quenched to room temperature. This treatment assured transformation of retained austenite. Tests showed that steels treated in this manner withstood substantially higher loads than those heat-treated conventionally. He pointed out, however, that merely attaining high hardness was no proof that a tool would function satisfactorily in the extrusion of steels. Of paramount importance is the ability of the steel to withstand alternating loads at temperatures up to 600 F.

Judging from published reports, the output produced by an individual tool seems to be subject to wide fluctuations. Part of the difficulty appears to result from excessive play in the extrusion presses; in a modern press, tools seem to fail from fatigue more often than from premature breakage or wear. A few reports of experience will indicate the differences in opinion among producers.

Morgan[11] reported difficulties with high-tungsten, high-speed types because the punches shattered frequently. The reason was not discovered but it may have been the result of inaccuracies in the press. At any rate, tooling was converted to a 2%-carbon, 12%-chromium steel. The life of the punch varied but production rates up to 8000 parts to the punch were obtained during the extrusion of low-carbon steels. He tried a high-silicon steel, which greatly improved the life of the tool, but when he finally converted the tools to a 5%-molybdenum, high-speed steel he was able to produce 60,000 parts before the punch failed.

TABLE 5:6
SUGGESTED COMPOSITIONS OF TOOL STEELS FOR EXTRUSION OF FERROUS METALS

Part	*Composition, %*								*Rockwell C Hardness*	*Source (Ref)*
	C	Mn	Si	Cr	Mo	V	W	Other		
Punches[a]	2.1	0.3	—	12	1	0.25	—	—	60-67	9
Punches[a]	1.6	0.4	—	13	1	0.25	—	—	58-60	
Punches[b]	0.8	—	—	4.2	5.0	2	6	—	64-68	
Punches[b]	0.75	—	—	4.0	—	1	18	—	64-66	
Die Rings	2.1	0.3	—	12	1	0.25	—	—	60-62	
Die Rings	1.6	0.4	—	13	1	0.25	—	—	58-60	
Die Bases	0.48	0.35	0.7	1.5	0.25	0.12	1.2	Ni 0.7	56-58	
Shrink Rings	0.34	0.60	0.35	1.4	0.4	—	—	Ni 4.25	44-48	
Punches	2	1.0	1.0	13	—	—	—	Ni 1.0	60-61	11
Punches[c]	—	—	—	—	5	—	—	—	—	
Punches	0.70	0.50	0.3	1.0	1.0	—	—	Cu 2.5	59-60	14

[a] For medium stressed parts
[b] For highly stressed parts
[c] Molybdenum high speed steel

Leland[12] states that punches and dies made of various high-speed steels and hardened to Rockwell C60-65 can be expected to produce from 25,000 to 50,000 parts before failure. If there is proper lubrication, failure is the result of fatigue rather than wear. Carbide tools have produced hundreds of thousands of parts without failure.

According to de Vore[13], punches for backward-extrusion, made of unspecified modified tool steels, have produced 300,000 parts from carbon-steel blanks without measurable wear. He noted also that solid tungsten carbide punches can be used satisfactorily in modern high-speed mechanical presses.

In general the compositions of the new tool steels have not been given in the publications in which they are mentioned. Riedel[14], however, mentioned a new tungsten-copper steel that was developed for cold-extrusion punches. This is an air-hardening steel with low distortion characteristics. It is particularly applicable for punches that fail by fracture or fatigue rather than wear.

Some suggested tool steels are given in Table 5:6.

REFERENCES

1. Tool Steels, *Steel Products Manual*, American Iron and Steel Institute, New York, 1955.
2. LANG, R. F., ANSELL, W. R., and WILSON, D. V., Impact Extrusion: German Practice, 1946, *B.I.O.S. Final Report No.* 1405, 1946.
3. GABLER, P., Precision Impact Extrusion and Die Casting of Aluminum in Industry, *Aluminium*, 25:194 (1943).
4. GALLOWAY, D. F., Some Experiments on the Impact Extrusion of Aluminum, Copper, and Steel, *Machinery* (London), 93:356 (1958).
5. BOSTON, J. R., Cold Impact Extrusion of Aluminum Parts for Douglas Aircraft, *Machinery*, 51, July 1945:138.
6. SHOEMAKER, J. D., Impact Extrusion of High Strength Aluminum, *Amer. Machinist*, July 6, 1953:129.
7. PATTON, T. L., Impact Extrusion of Magnesium, Paper delivered at the 7th Annual Meeting of the Magnesium Association, 1951.
8. SABROFF, A. M., SANNICANDRO, R. A., and FROST, P. D., Design Considerations for Cold Extrusion of Titanium, Paper No. 119, *American Society of Tool Engineers*, May 1958.
9. HOWARD, F., DENNISON, H. A. J., and ANGUS, N., Cold Extrusion of Steel, *Sheet Metal Industries*, 38:403 (1961).
10. JOHNSON, E., and BISHOP, E., Die Steels for Cold Extrusion, *Ibid.*, 30:490 (1953).
11. MORGAN, R. A. P., The Cold Extrusion of Steel, *Metal Treatment*, 28:393 (1961).
12. LELAND, J. F., Tooling for Cold Steel Extrusion, *Machinery*, July 1955:170.
13. DE VORE, W., Cold Extrusion of Steel, *Sheet Metal Industries*, March 1954:197.
14. RIEDEL, J. Y., Air-Hardening Tool Steels Offer Lower Costs, Improve Performance, *Metal Progress*, May 1960:98.

Chapter 6

LUBRICATION

The type of lubrication required for impact extrusion is different from that required for bearings. Movement is slower in metal-forming, pressures are higher, and it is not so easy to force the lubricant into the contact area.

Under the high pressures encountered, atoms on the surface of the metals tend to weld on contact. To prevent this contact, a lubricant must not only form a protective layer but also lower the friction between the surfaces. Even in the finest polished surfaces, there are peaks of surface roughness which tend to touch and break the film on the surface which cannot be renewed.

Suitable lubrication of soft metals for impact extrusion is not a great problem. As the strength of the metals being processed increases, however, the problem of lubrication becomes more serious. Additives to oils must be more active as the pressure increases and these additives lead to the use of *extreme-pressure lubricants.*

In a discussion of the action of such lubricants, Williams[1] points out that the capacity of a lubricant to withstand the service requirements can be indicated by the ratio

$$\frac{T_c - T_i}{\mu}$$

where T_c = critical temperature of the lubricant
T_i = bulk temperature of the metals in contact
μ = coefficient of friction at the contact point

Load-carrying capacity can be increased by raising the critical temperature, T_c, at which the lubricant breaks down, or by reducing the coefficient of friction.

Extreme-pressure lubricants can be formed by adding chlorinated hydrocarbons to a lubricating oil or grease base. Under conditions of light loading, these compounds are inert to most metals. As the pressure increases, local heating occurs at incipient local welding, the compound reacts with the metal to form a chloride at the spot, and protects it. Most metallic chlorides are solids that are easily sheared below their melting points and thus act as lubricants under extreme pressures. Some metallic sulfides and oxides also have these properties.

Extreme-pressure lubricants have been used successfully for the impact extrusion of some metals. When the process was applied to steel, however, it was found that a preliminary treatment of the surface was required. Generally, this consisted in coating it with zinc phosphate and, although some producers use extreme-pressure lubricants with the phosphated surface, the use of soaps which react to form insoluble zinc soaps is more common. These lubricants will be discussed in another section.

The lubricant also affects the finish of the surface of the extruded item. Butler[2] showed that surface finish deteriorates with increasing viscosity of the lubricant. He suggests that lubricant trapped at the interface is a factor that affects both the frictional resistance at the tool metal interface and the surface finish that results. These trapped films may be formed mechanically at surface depressions or by hydrodynamic mechanisms. He states that the support given by these films combined with the separation of tool and metal at the interface explain the low coefficients of friction and the surface finish.

To obtain a polished finish on the part, the tool and metal must conform without metallic pick-up or scoring. According to Butler,[3] the lubricant can influence this action. He notes that a thin semi-continuous film may be easily trapped at the interface and, as a result, the metal is deformed largely through fluid pressure. If this condition occurs, a matte non-reflecting finish is produced on the deformed surface.

These general comments on lubrication and its effect on the extruded part must be supplemented by specific information on lubrication for particular metals. There is no universal lubricant for impact extrusion.

SELECTION OF LUBRICANTS

The selection of lubricants for specific metals is empirical. A lubricant that is perfectly satisfactory for lead may or may not be suitable for aluminum, and probably will be totally unfit for the extrusion of steel. There are, however, certain guideposts to lead toward proper selection.

Qualitative indications can be obtained by several methods. It requires a higher pressure to extrude a part with use of a poor lubricant than with a good one. In a mechanical press, this difference makes itself apparent by a distinct difference in the sound of pounding of the press. Boston[4] points out that the effect of lubrication can be determined by measuring the height of the shell. With a poor lubricant, he found wide variations in height, whereas with a good lubricant, variation was much less pronounced. Finally, assuming that the tools are in proper condition, the surface finish of a part produced with a good lubricant will be much better than that produced with a poor one.

As the lubricant is usually selected after a series of tests, it is a subject that is considered a plant secret by some producers and they refuse to discuss lubrication. To indicate suitable lubricants, therefore, it is necessary to refer to a limited number of publications. This is not intended to imply that these lubricants are the only suitable ones for use with the metals mentioned, but rather to suggest them as starting points for development. Variations in individual plants, in tooling, or in the part being extruded, may make one lubricant preferable to another.

In discussing lubrication, the topic will be broken down into lubrication of untreated surfaces and lubrication of treated surfaces.

LUBRICATION OF UNTREATED SURFACES

The first application of impact extrusion was to the soft metals, lead and tin, and it was not particularly difficult to find lubricants that could be applied to the untreated surface. As the process was extended to other nonferrous metals, lubrication of the untreated surface was continued and satisfactory lubricants were developed. When attempts were made to extrude steels, however, they failed because the lubrication was unsatisfactory. Only by treating the surface with a coating that could hold the lubricant was it possible to extend the process to the ferrous metals. Surface treatments before lubrication have been applied also to some of the stronger nonferrous metals. This subject will be discussed later.

Lead and Tin

The problem of lubricating lead and tin alloys for impact extrusion are less serious than they are for other metals. Lead acts as a lubricant and, to a lesser extent, tin does also. In addition, both metals are soft and they are actually being hot-worked at the temperatures reached during commercial extrusion. As one producer expressed it, lead can be backward extruded even with water as the lubricant.

One of the best lubricants for these metals is zinc stearate. As the residue is difficult to remove and is incompatible with coating materials, this lubricant can be used only for parts that do not require decoration.

A survey of producers revealed that a number of lubricants were being used, selection being based on compatibility with the coating and decorating materials that were being used in the specific plant. Among these lubricants were light mineral oil, petroleum jelly, vegetable shortening and talc, olive oil and cornstarch, castor oil and talc, and castile soap. Various proprietary greases have also been used successfully.

The lubricant is generally applied by tumbling blanks for a short time in a standard tumbling barrel. The quantity of lubricant should be the minimum that will serve; excessive quantities can cause defects in the walls of the extruded part and can interfere with the action of the ejecting mechanism of the press.

Zinc and Zinc Alloys

Little information has been published on the impact extrusion of zinc parts. Although the crystalline form of this metal is less ductile than that of lead, it recrystallizes at a low temperature and therefore undergoes hot-working during commercial extrusion. Mabb[5] notes that lanolin is satisfactory for lubrication in the backward extrusion of parts from high-purity zinc.

Aluminum and Aluminum Alloys

The lubrication problems in the extruding of aluminum are more serious than they are with lead and tin. The metal is harder and it has tendencies toward galling and seizing that must be overcome. A number of lubricants have been suggested that permit satisfactory production operations, and a favorite constituent of these lubricants is lanolin.

Shoemaker[6] states that lubricants must have high film strength and flash points above 500 F because of the high temperatures reached by the part during extrusion. Lubricants based on animal fats such as lanolin and lard oil, or on their compounds, are satisfactory for aluminum. He notes that although certain types of waxes are applicable they are not recommended because excessive pick-up of wax galls the surface of the part being extruded. The residual wax in the die congeals and this interferes with the ejecting action of the press. He suggests that a thin film of lubricant is generally all that is required but as the thickness of the film will vary with the size and shape of the part this must be determined by experiment. The lubricant should be diluted to the proper consistency with volatile solvents such as trichlorethylene or carbon tetrachloride, and applied by brushing, dipping, or spraying. The solvent should be permitted to evaporate before extrusion of the part.

For the commercial production of aluminum cans in Germany,[7] the following mixture was used:

Talc	0.9 kg
Cylinder oil	1.1 kg
Animal fat	0.2 kg
Glycerol	0.03 kg
Carbon tetrachloride	10 liters

This lubricant is applied by tumbling for a short time and the lubricated blanks are stored for two days before extrusion.

Galloway[8] suggests tumbling the blanks for a short time in a 20 per cent solution of lanolin in light oil. Pineau[9] used a mixture of tallow and lanolin dissolved in carbon tetrachloride. He stated that the film remaining after the solvent evaporated admirably fulfilled the requirements of minimum extrusion pressure and good surface finish. Various proprietary extreme pressure lubricants have been used also for the extrusion of aluminum.

In general, the recommended procedure for lubrication is to use as light a film as possible and most producers suggest tumbling for a short time in a standard tumbling barrel. Too much tumbling, however, results in the formation of aluminum powder which may adhere to the blank and cause pin-holing during the extrusion of thin-walled components.

Magnesium and Magnesium Alloys

Impact extrusions of magnesium are produced at moderately elevated temperatures and consequently the lubrication problem is somewhat different from that encountered with the materials discussed previously. Although many lubricants have been tried, the only successful ones appear to be dispersions of graphite. According to Patton[10] the lubricant should be applied by tumbling or spraying. Only a small quantity is required—about one cubic centimeter for each 100 square inches of surface. If the coating is too heavy, it causes the part to stick in the die.

Copper and Copper Alloys

Little information has been published on the impact extrusion of copper and its alloys. Galloway[8] recommends using a mineral oil of medium viscosity as a lubricant for backward extrusion. Landberg[11] states that in the extrusion of commercial copper the best results are obtained with a calcium stearate lubricant.

For the forward extrusion of copper parts, Crampton[12] notes that jets of lard oil played continuously on the die furnished satisfactory lubrication.

LUBRICATION OF TREATED SURFACES

Proper preparation of the surface is essential for the successful cold extrusion of steel. It appears to be required also for some of the stronger nonferrous alloys.

Carbon and Low-Alloy Steels

For carbon and low-alloy steels, the surface treatment consists of forming a phosphate coating. Formation of this coating requires three distinct steps; cleaning, phosphating, and lubricating.

As the first step, it is essential that the surface be completely free of grease, rust, and mill scale. The phosphate coating can be formed only on an absolutely clean surface.

Phosphating consists of treating the clean surface of the metal with a solution containing primary iron, manganese, or zinc phosphate, free phosphoric acid, and frequently an accelerator to reduce the time required for the coating to form. A coating of zinc phosphate is almost always used for cold extrusion. Details of the coating procedures have been published by Holden,[13] Pimbley,[14] Leland and Helms,[15] James,[16] and others.

Holden[13] states that a phosphate can be formed readily on carbon steels and on a wide variety of low-alloy steels. Under equal processing conditions, thicker coatings are formed on high-carbon than on low-carbon steels. This coating forms only with difficulty on steels that contain more than five per cent chromium, and this treatment cannot be used for the austenitic chromium-nickel steels. For heavy reductions of area, the coating of zinc phosphate should weigh at least 1000 milligrams to the square foot.

The phosphate coatings are porous, crystalline, and insoluble in water. They adhere tenaciously to the surface of the metal and have a high absorptive capacity for organic compounds such as oils and waxes. The final treatment consists of reacting these coatings with suitable compounds.

Experimental work has shown that lubricants containing polar groups give excellent results with phosphate-coated surfaces. Such polar lubricants as soaps, vegetable oils, and animal fats are suitable for use with zinc phosphate coatings, particularly if they are applied in hot solutions. These compounds react with the phosphate to form insoluble zinc soaps.

Holden[13] points out that the composition, concentration, pH, temperature, and method of application of the lubricants influence the amount and rate of formation of the lubricating film. When soaps, such as sodium stearate, are dissolved in water, they hydrolyze into sodium hydroxide and stearic acid or acid soap, the degree depending on the temperature and the dilution. The free acid or the acid soap is absorbed by the phosphate on the surface. The sodium hydroxide reacts with the zinc phosphate in the coating, releasing zinc ions which, in turn, react with the stearate ions to form insoluble zinc stearate. This stearate precipitates within the pores of the coating; therefore the lubricating film may be considered to be formed by a combination of adsorption and chemical reaction.

An experimental investigation showed that only five per cent of the film formed on a zinc phosphate coating immersed in a sodium stearate solution was soluble in water and this consisted of sodium stearate or acid soap. The remainder was zinc stearate and was soluble only in organic solvents.[17]

TABLE 6:1

EFFECT OF VARIOUS PHOSPHATE COATINGS ON EXTRUSION PRESSURE FOR A 0.10% CARBON STEEL*

Coating Material	*Lubricant*	*Maximum Load psi*
Parkerizing B	Bonderlube 235	296,000
Bonderite 181X	Bonderlube 235	269,000
Bonderite 160X	Bonderlube 235	284,000
Bonderite 165X	Bonderlube 235	283,000
Aluminum Phosphate	Bonderlube 235	296,000
Parkerizing B	Palm Oil Emulsion	274,000
Bonderite 181X	Palm Oil Emulsion	272,000
Bonderite 160X	Palm Oil Emulsion	283,000
Bonderite 165X	Palm Oil Emulsion	269,000
Aluminum Phosphate	Palm Oil Emulsion	287,000

* Kunogi[22]

Kunogi[22] investigated the effects of changing the type of coating on the extrusion pressure required for a 0.10%-carbon steel, using proprietary coating materials. Some of his data are given in Table 6:1. He stated that zinc phosphate coatings were the best for steels.

TABLE 6:2

EFFECT OF VARIOUS LUBRICANTS ON EXTRUSION PRESSURE FOR A 0.22% CARBON STEEL*

Coating Material	*Lubricant*	*Maximum Load psi*
Bonderite 181X	Bonderlube 235	305,000
Bonderite 181X	Special Fatty Acid Soap	284,000
Bonderite 181X	Stearic Acid	290,000
Bonderite 181X	Calcium Stearate	306,000
Bonderite 181X	Aluminum Stearate	294,000
Bonderite 181X	Palm Oil Emulsion	292,000
Bonderite 181X	Colloidal Graphite	310,000

* Kunogi[22]

He also investigated the effects of changing the lubricant on the extrusion pressure required for the extrusion of a 0.22%-carbon steel. Some of these data are given in Table 6:2. The lowest pressure resulted from the use of a fatty acid soap, the highest from colloidal graphite.

SPECIAL CONSIDERATIONS

Having discussed the coating problem in general terms, it appears desirable to give some specific examples of use. A survey of producers of cold-extruded steel parts indicated that all were using a zinc phosphate coating as the base. Several were treating phosphate-coated carbon- and low-alloy steels that contained up to 8 per cent chromium with a high-titre sodium soap to form insoluble zinc soaps; one was using molybdenum disulfide as the lubricant for the extrusion of carbon steels.

Sieber[18] suggested soaking the phosphate-coated carbon steel blank in a warm soap solution rich in fatty acids. He noted that mineral oil was not a satisfactory lubricant because it did not react with the phosphate coating. By contrast, Galloway[8] recommended forming a phosphate coating by immersing the steel blank for sufficient time to produce a coating in the thickness of 1800 milligrams to the square foot, and dipping the blank into an extreme-pressure lubricant just before extrusion.

Morgan[19] recommends a coating of zinc phosphate weighing 2000 milligrams to the square foot, and a proprietary activated soap solution for the lubricant. The blank is immersed in the lubricant at 160 F for five minutes to permit formation of a thin, even coating; too thick a coating fouls the die. He noted that, although this system of lubrication is satisfactory, the smooth surface of the component is improved if the billet is dusted with a powdered extreme-pressure lubricant. The disadvantage of using this powdered lubricant is that it tends to remain in the die cavity and become packed solidly on the bottom.

Leland and Helms[15] recommend coating with phosphate by immersion in an acid zinc phosphate solution at 160 to 170 F. They do not agree with others who state that weight of coating is a criterion for judging the suitability of the zinc phosphate coating. The blank is lubricated by immersion in a dilute reactive soap solution at 140 to 160 F for three to seven minutes to form a water-insoluble zinc soap.

A number of other articles have been published that deal with phosphate coating and lubrication for the cold extrusion of steel but the suggestions are essentially the same as those cited.

Stainless Steels

Holden[13] notes that zinc phosphate coatings do not adhere to the austenitic stainless steels and that difficulties are encountered with straight chromium steels containing more than 8 per cent of chromium. The austenitic chromium-nickel stainless steels, of which the 18%-chromium, 8%-nickel type is the most familiar, can be coated with oxalate, and lubricated in a manner similar to that described for the phosphate-coating of carbon steels.

Eshleman[20] states that Type 410 chromium stainless steel, which contains 12 per cent of chromium, can also be treated to form an oxalate coating. A complex soap lubricant is used with this coating.

Nonferrous Metals

Little information has been published on the use of phosphate or other coatings for the impact extrusion of nonferrous metals. The author compared the pressures required to extrude commercial aluminum blanks lubricated conventionally with others that had been coated with phosphate before lubrication. The extrusion pressure required for the phosphate-coated blanks was approximately ten per cent lower than that required for blanks lubricated in the conventional manner.

In one survey, a producer reported using a phosphate coating with a soapy lubricant for the backward extrusion of the aluminum alloys 3003,2024,6061,6063, and 7075.

Sabroff and Frost[21] developed a fluoride-phosphate coating which was used successfully for the cold extrusion of titanium. They used a self-drying gum-resin mixture, containing graphite and molybdenum disulfide as the lubricant.

It would seem that suitable surface treatments could be developed for the extrusion of other strong nonferrous alloys, but no data have been reported.

REFERENCES

1. Williams, C. G., Mechanism of the Action of Extreme Pressure Lubricants, *Proc. Royal Society*, 212A, 1952, 512.
2. Butler, L. H., The Effect of Lubricants on the Growth of Surface Contact Areas During Plastic Deformation of Metals, *J. Inst. Metals*, 89:116 (1960).

3. BUTLER, L. H., Surface Conformation of Metals Under High Nominal Contact Pressure, *Metallurgia*, 61, April 1961:167.
4. BOSTON, J. R., Cold Impact Extrusion of Aluminum Parts for Douglas Aircraft, *Machinery*, 51, July 1945:138.
5. MABB, P., Impact Extrusion, *Machinery* (London), February 9, 1950:192.
6. SHOEMAKER, J. D., Impact Extrusion of High Strength Aluminum, *American Machinist*, July 6, 1953:129.
7. Impact Extrusion in Germany, *Machinery* (London), October 23, 1947:460.
8. GALLOWAY, D. F., Some Experiments on the Impact Extrusion of Aluminum, Copper, and Steel, *Ibid.*, 93, August 13, 1958:356.
9. PINEAU, M., Backward Impact Extrusion, *La Pratique des Industries Mecaniques*, 37, June 1954:163.
10. PATTON, T. L., Impact Extrusion of Magnesium, Seventh Annual Meeting of Magnesium Association, 1951.
11. LANDBERG, P., Some Remarks on Impact Extrusion, *Microtechnic*, 12, December 1958:301.
12. CRAMPTON, D. K., Extrusion of Metals, *Trans. Amer. Soc. Metals*, 25:55 (1937).
13. HOLDEN, H. A., Phosphate Coatings for Assisting Cold Extrusion, *Sheet Metal Industries*, June 1953:502.
14. PIMBLEY, G. H., How to Get the Most from Phosphating Solutions, *Metal Progress*, August 1961: 103; September 1961: 108; November 1961:108.
15. LELAND, J. F. and HELMS, J. W., Troubles with Cold Extrusion May Be Traced to Lubrication, *SAE Journal*, November 1954: 43.
16. JAMES, D., Phosphate Coating and Lubricating Steel for Cold Extrusion, *Sheet Metal Industries*, March 1961:171.
17. *Cold Shaping of Steel—Summary Report*, Office of Technical Services, U.S. Department of Commerce, PB96704, 1947.
18. SIEBER, K., Developments in Cold-Flow Pressing and Extrusion of Steel, *Machinery* (London), 80:891 (1952).
19. MORGAN, R. A. P., The Cold Extrusion of Steel, *Metal Treatment*, 28, October 1961:393.
20. ESHLEMAN, R. H., Cold Extrude Stainless Shafts, *Iron Age*, January 21, 1960:86.
21. SABROFF, A. M., and FROST, P. D., A Comparison of Lubricants and Coatings for Cold Extruding Titanium, *ASLE Trans.*, 3, April 1960:61.
22. KUNOGI, M., Cold Extrusion of Alloy Steels, *Japan Soc. Mechanical Engineers*, 23, October 1957:249.

Chapter 7

HIGH ENERGY RATE FORMING

During the past few years there has been a revival of interest in working metals at high velocities. Although this development, called *high energy rate forming*, is not new it has recently been expanded into new directions. The novelty lies in its application to operations usually performed in presses.

A comparison with the " high velocity " of conventional equipment will indicate the differences in forming rates. Whereas some high speed presses operate at a velocity of 800 feet per minute, high energy rate forming speeds of 20,000 feet per minute are possible.

In attempts to explain the reasons for the ability to form metals at high velocity, history is repeating itself. In the early days of development of impact extrusion, the ability to form a part was attributed to the impact velocity with which the punch struck the blank. It was considered that a finite interval of time elapsed before the material work-hardened and the speed permitted flow to occur during this interval. Similar theories are being advanced now to explain forming by high energy rates. In the case of impact extrusion, further work has convinced most scientists and engineers that impact is not essential and that a steady pressure will accomplish the same result.

Some of the scientists working with high energy rate forming believe that greater elongation results when the materials are worked at high velocities. Others believe that forming at high energy rates follows the usual rules, the major difference lying in the time during which the forces are applied. With the background of the developments in impact extrusion as a guide, one may venture to predict that the latter theory will prove to be correct.

Regardless of theories, however, considerable experimental work has shown that a number of metals, difficult to form by conventional methods, can be worked by high energy rate processes. This is another example of a process in which practical experience has outstripped theory.

Brown[1] posed several pertinent questions concerning high energy rate forming. Particularly interesting are:

> How many parts produced by the process could be made as easily by slight variations of established processes?
>
> How many parts produced by the method today are simply " last resort " types of thinking; and how many have been actually designed for this type of production right from the beginning?

So much of the work that is being done is held under restrictions of secrecy that these questions remained unanswered. From the published literature, however, it is possible to describe in general terms the methods of performing the process.

Several methods of high energy rate forming are in operation today in various stages of development. They are:

1. By gas
2. By explosion
3. By electrical methods
4. By magnetic methods

These methods, some of which may be applicable to backward or forward extrusion, are discussed briefly in the following sections.

FORMING BY GAS

In the method of forming by gas, compressed gas is released to accelerate a piston to a high velocity. The piston strikes a workpiece and forms it into the desired shape. The machines can be used for forging, forming, backward extrusion and forward extrusion. Two types of commercial equipment that have been produced operate on this principle. They are the Dynapak and the Clearing Hermes.

The Dynapak Machine

Basically, the Dynapak consists of two chambers separated by a plate with an orifice. A piston in one chamber is attached to a column that acts as the moving member of the tooling combination. The second chamber is a reservoir for storing nitrogen at high pressure. A small sealing ring or gasket is attached to the piston head to surround the opening in the orifice plate. The machine is shown diagrammatically in Figure 7:1.

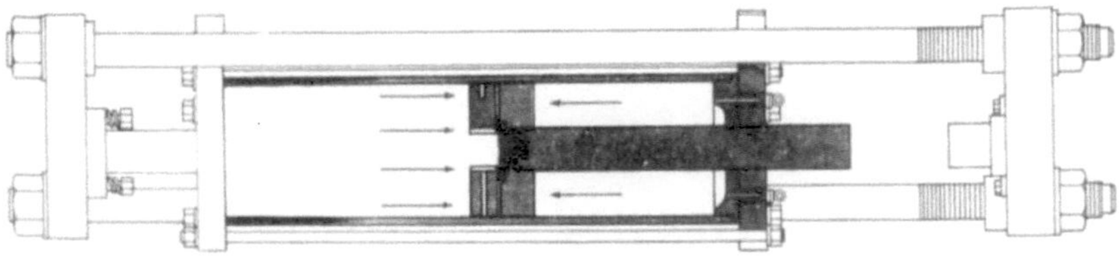

FIG. 7:1. DIAGRAMATIC SKETCH OF DYNAPAK MACHINE (General Dynamics Corporation)

If the pressure is raised in the chamber containing the piston, the piston head is forced against the orifice plate and the seal isolates all the area of the piston head, excepting that lying within the sealing ring, from the second chamber. As the pressure in the storage chamber can act only on the area of the piston head, which lies within the sealing ring, a high pressure must be produced in this chamber to effect a balanced thrust on the piston.

For example, if the area of the piston is 25 square inches and the area of the orifice is $2\frac{1}{2}$ square inches, the ratio between the two is 10:1. Now, if the sealing ring on the piston is forced against the orifice plate with a gas pressure of 100 *psi*, the pressure in the storage chamber must be 1000 *psi* to balance this force. This differential is required because the pressure in the storage chamber is applied only on $2\frac{1}{2}$ square inches whereas the pressure in the other chamber is applied over the entire piston area of 25 square inches. Under these conditions, the system is in equilibrium. To upset this equilibrium, the pressure in the high pressure chamber must be raised slightly. This additional pressure starts movement of the piston, disengaging the seal around the orifice and permitting the high pressure to be applied to the entire face of the piston. As a result, the low pressure in the piston chamber is greatly overbalanced and the piston is driven forward at speeds as high as 2500 feet a second. The actual speed depends on the size of the cylinders and the pressure differential in the two chambers. The moving piston strikes the workpiece with energy levels as high as 1,500,000 foot-pounds. The process has been described in some detail by Fedderson.[2]

The Dynapak has been used to produce a number of parts by cold extrusion. An example is a gun bolt produced from a bar of C-1118 carbon steel. The billet was lubricated with a soap-type lubricant before extrusion and was cold-extruded, using a firing pressure of 900 *psi*. The only finishing required was trimming the part to length.

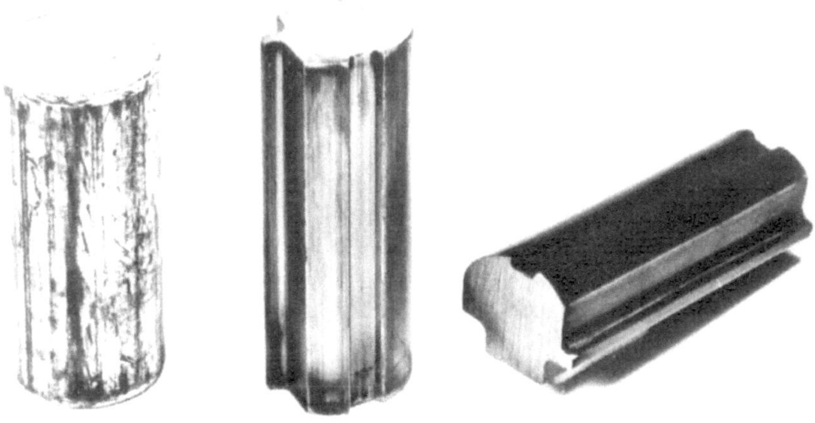

FIG. 7:2. GUN BOLT SLIDE, COLD-EXTRUDED ON DYNAPAK FROM COLD-DRAWN BAR OF C-1118 STEEL

This part could be produced on a conventional cold-extrusion press, but this shows that the Dynapak can be used successfully for cold extrusion. Development is in too early a stage to permit prediction of the possibility of using this machine for the cold extrusion of the high strength metals that are difficult to extrude by conventional methods.

The Clearing Hermes Machine

This machine operates also on compressed nitrogen and the basic principle is similar to that described for the Dynapak. It employs two pistons, however, and two platens are moved towards each other at high velocity. A description of one model notes that two 2.5-ton platens are moved toward each other at a velocity exceeding 60 miles per hour.[3] The machine is designed for mass production and it can produce six finished parts per minute. Although designed as a forging press, it appears to be adaptable to the production of parts by cold extrusion.

FORMING BY EXPLOSION

In forming by explosion the explosive acts merely as a source of energy and is comparable in its effect with the action of a mechanical or hydraulic press. The energy, however, is highly concentrated.

Explosives can be divided into two classes, low and high. Typical examples of low explosives are black and smokeless powders. Typical high explosives are TNT, RDX, and picric acid. The difference between the two types lies in their mode of action.

A low explosive deflagrates, or burns, producing relatively low pressures in the order of 40,000 *psi* at a conversion time measured in thousandths of a second. *Conversion time* is the time required to transform a working quantity of the explosive into gaseous products. A high explosive detonates, producing pressures in the order of 2,000,000 *psi* or greater, at a conversion time measured in millionths of a second.

The energy of an explosive is not released instantaneously. There is a finite interval of time during which the material is converted into gaseous products. The velocity with which this conversion front moves through the explosive is called the *deflagration velocity* for low explosives and the *detonation velocity* for high explosives. For low explosives, this velocity is measured in inches to the second, or in a few feet to the second; for high explosives, it ranges up to 28,000 feet a second.

In most contemporary work, high explosives are used, but pressures of several million *psi* are too high to be used successfully for most metal-working operations. Lower pressures are achieved by placing the explosive charge at a distance from the workpiece and transmitting the energy through air or water. This is known as a *stand-off operation.* By selecting the type and amount of explosive, the shape of the charge, the transmitting medium, and the distance, it is possible to obtain working pressures that vary from a few thousand to hundreds of thousands of pounds.

The second type of operation is the *contact* type, which as the name implies, is performed with the explosive in direct contact with the workpiece. In this type of operation, much of the energy is transmitted directly to the metal as a stress or shock wave of high intensity.

Explosive forming is being used or developed for a variety of metal-working operations. Those employing stand-off charges include forming, flanging, and sizing of flat products. Those that utilize contact charges include welding, compacting of metal powder, controlled work-hardening, extruding, and cutting.

The major advantages of explosive forming are that usually only very small quantities of the explosive are required, and that the volume is small. Therefore the operation can be performed at the site of the work instead of bringing the workpiece to a press or other machine.

Other advantages include the low cost of equipment required; the ability to form pieces so large that they are beyond the capacity of most presses; and the possibility of producing a limited number of parts without the high tooling charges that would be required to produce the same parts on conventional equipment.

Explosive forming offers no advantage for long production runs or for parts that can be produced readily on conventional equipment. Therefore it is doubtful whether the process will ever be employed extensively in impact extrusion although it might serve for the experimental production of parts in limited quantities.

Explosive forming is still in the early stages of commercial exploitation and very little information is being released by the companies that are working with the process. The procedures have been discussed by Pearson,[4] Tardif,[5] and Simons.[6]

FORMING BY ELECTRICAL METHODS

High energy rate forming can also be performed by electrical methods. These methods are also explosive in nature, the explosion being produced by a spark that jumps between two electrodes or by exploding a fine wire connecting the electrodes. The discharge is usually produced under water and the workpiece is shaped by the shock wave resulting from the explosion.

High voltages at high capacities are required. According to Fedderson,[2] a voltage of 4000 with a condenser capacity of 1200 microfarads can be used. The energy contained in the condenser bank equals about 11,000 foot-pounds. When the condensers are discharged, an increase in pressure of approximately 15,000 atmospheres occurs and the shock wave is transmitted through the water to the workpiece.

High energy spark forming is suitable for expanding and sizing hollow bodies in particular and this appears to be the most economical application. This procedure does not appear to offer any possibility for future use in impact extrusion.

FORMING BY MAGNETIC METHODS

For forming by magnetic methods an insulated induction coil is wrapped around or placed inside the workpiece. An electrical charge drawn from a bank of capacitors sets up opposing currents in the coil and in the metal that is being worked. These currents produce a strong magnetic force that deforms the metal.

This method is just being developed. It appears to be suitable, primarily, for tubular parts, but to have no application in impact extrusion.

The underlying principles, advantages, and disadvantages of magnetic forming have been discussed by Wagner and Boulger.[7]

REFERENCES

1. Brown, H., New Approaches to Metal Fabrication, *Battelle Technical Review*, 11, March 1962.
2. Fedderson, E. W., High Energy Rate. . . Its Application to Metal Forming, *Journal of Metals*, September 1960:682.
3. New One-Step Forging Machine Presses for Higher Outputs, *Iron Age*, January 25, 1962:100.
4. Pearson, J., Metal Working with Explosives, *Journal of Metals*, September 1960:673.
5. Tardef, H. F., Shape, Fasten, Engrave, Test Materials with Explosives, *Materials in Design Engineering*, February 1959:82.
6. Simons, C. C., Explosive Metalworking, *DMIC Memo No. 71*, Battelle Memorial Institute, November 3, 1960 (OTS, PB 161221).
7. Wagner, H. J. and Boulger, F. W., High Velocity Metalworking Processes Based on Sudden Release of Electrical Energy, *DMIC Memo No. 70*, Battelle Memorial Institute, October 27, 1960 (OTS, PB 161220).

Part 2

THE PRODUCT

Chapter 8

FACTORS IN DESIGN

Before designing for impact extrusion, many factors should be considered. As has been mentioned previously, there are certain items that can be made only by this process, for example, collapsible tubes. As the greatest number of parts can be produced by more than one method, selection of the most practical one is often a matter of economics.

The use of impact extrusion should be considered when:

1. The length of the part is greater than two to three times the diameter.
2. The part is hollow with one end closed or partially closed.
3. The part would require considerable machining if produced by some other method.
4. The base must be thicker than the side walls.
5. Zero draft is desired in the side walls.
6. Redesign will permit use of one piece instead of an assembly.
7. Large quantity production is planned.

Impact extrusion should probably not be considered if:

1. The part can be made easily on a screw machine.
2. Only a few parts are required.
3. The part is unsymmetrical and the length of it is greater than three to four times the diameter.
4. The bottom must be flat from edge to edge.

If after careful analysis the process seems to offer possibilities for economical production of the part by impact extrusion, proper design should be developed. Requirements for the design of impact-extruded parts are considered in the following sections.

BACKWARD EXTRUSION

In the production of parts by backward extrusion, the punch must be rather long, and as it is supported only at one end it has a tendency to float. For this reason, it is desirable to aim for symmetry in the design of parts.

Figure 8:1 shows a few examples of good and poor design. Parts on the left can be produced with close tolerances and with minimum lateral movement of the punch. In producing the

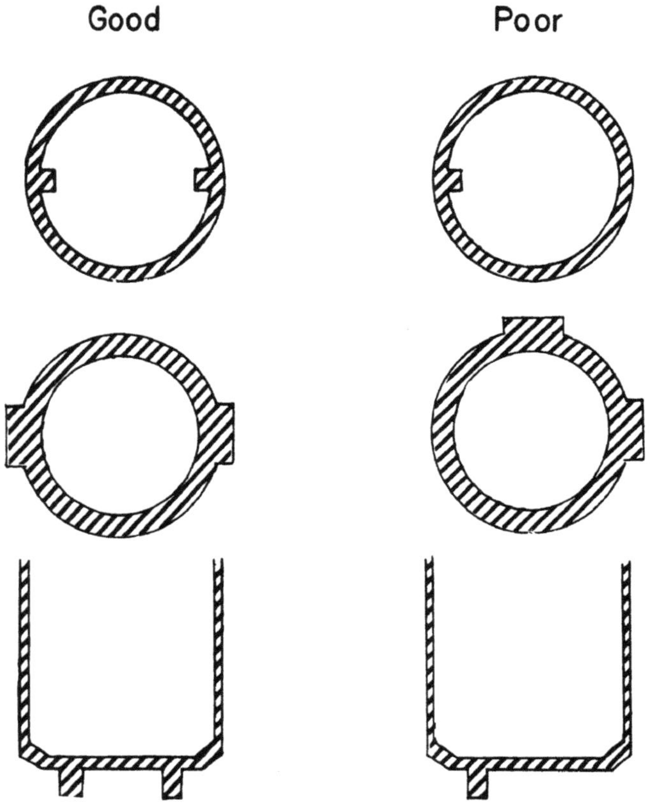

FIG. 8:1. SOME EXAMPLES OF GOOD AND POOR DESIGN

parts shown on the right, the forces tend to be off-center and lateral movement can occur. One way to minimize floating of the punch is to include a small recess in the design, as shown in Figure 8:2; the point on the tip of the punch required to produce this recess helps to maintain alignment.

In backward extrusion, the metal is forced to flow through an orifice to form the sidewalls; therefore, parts must be so designed that there will be no restriction to flow. There should be no sharp angles at points where the flow of metal changes direction. Instead, as shown in Figure 8:2, flow should be assisted by incorporating suitable punch angles and by rounding the corners of the die. The ideal design for the bottom, to ensure minimum restriction to flow, incorporates an angle of 120

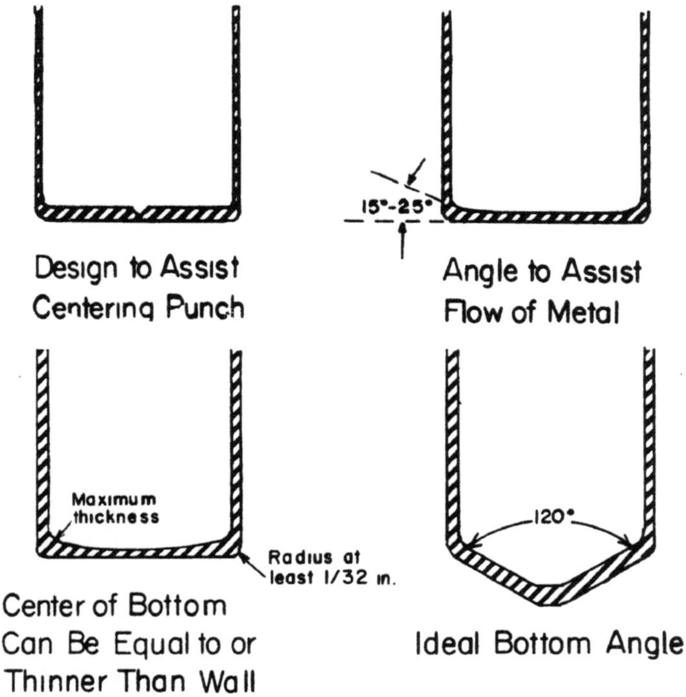

FIG. 8:2. SOME POINTERS ON DESIGN

degrees at the bottom, as shown. Incidentally, in the earliest commercial product, the collapsible tube, the angle at the shoulder is approximately 120 degrees.

The thickness of the bottom of a backward extrusion is independent of the thickness of the wall. This fact gives the impact-extruded part distinct advantage over a deep-drawn part in some applications. The bottom can be thicker than the walls or equal in thickness. It can also be tapered to the center with this area thinner than the walls. In all designs, however, the thickest part of the bottom should be at the junction of the bottom with the wall, as shown in Figure 8:2.

The length of a part formed by backward extrusion is limited by the strength of the column of the punch. This subject was discussed under tooling. For commercial operations in general, the length should not be greater than six to eight times the diameter, although greater ratios have been used in producing experimental extrusions. The greater the length, the greater is

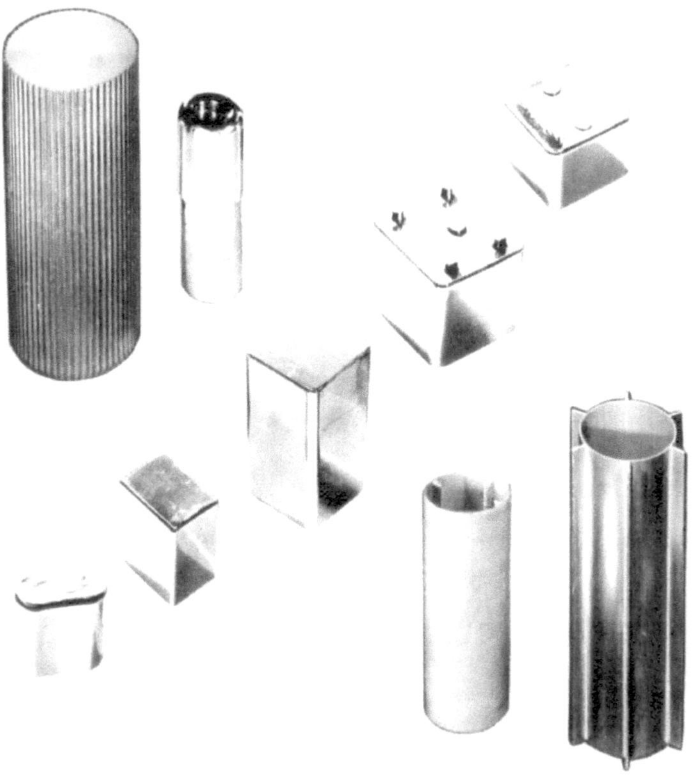

FIG. 8:3. TYPICAL BOTTOM CONTOURS, BOSSES, AND RIBBED SHELLS

the tendency of the punch to float, and consequently the greater the variation to be expected in the thickness of the wall.

This chapter is concerned only with the general features of design, and therefore it will not go into minimum wall-thickness, maximum ratio of length to diameter, and other factors that vary with the metal being extruded. Points of this kind are considered under the various materials discussed in the following chapters. For the same reason, only typical parts are discussed and illustrated in this chapter. Specific examples are given in succeeding chapters.

Other features that can be incorporated into the design of parts for backward extrusion are:

Bottom Contours

The great majority of shells have circular cross sections but this shape is not a requirement. Bottoms can be circular, oval, square, rectangular, or other regular polygons. They can also be irregular in cross section as long as they are symmetrical. Some forms are shown in Figure 8:3.

Bosses

Internal and external bosses can be produced as integral parts of the design. Internal bosses can be longer than the side-walls, if desired. External bosses can be produced in any size up to the diameter of the shell; they can be solid or hollow. Internal and external bosses can be produced on the same shell. They can be of the same or different diameters and lengths and of different cross sections. These cross sections can be round, oval, square, or of any other regular configuration. Some types of bosses are shown in Figure 8:3.

Ribs

Longitudinal ribs can be produced on either the inside or the outside of the side walls. Outside ribs can be spaced equally or arranged in a symmetrical design. They may serve either as functional parts of the design or be added for decoration. Internal ribs can be designed to run the full length of the shell or to stop at a desired distance from the closed end. Inside ribs often serve to reinforce the thickness of the wall. Some examples of ribs are shown in Figure 8:3.

Multi-Wall Shells

Double-wall shells can be produced readily. The diameter of the center tube can be varied to provide a greater or lesser opening in the center than between the internal and external walls. The center tube can also be longer than the outside walls, if desired. A double-walled extrusion is shown in Figure 8:4. The process is not limited to double walls, however. Multi-wall extrusions are also produced, and examples of the use of such parts are discussed in Chapter 9.

Unsymmetrical Parts

Although it is preferable to design symmetrical parts, especially for the production of shells having large ratios of length to diameter, unsymmetrical parts can also be produced if the height is not too great. For the production of such items, it is necessary to ensure uniform flow of metal to all corners of the die by adjusting the angles of the tools. This procedure is strictly on a cut-and-try basis and may be quite time-consuming.

FIG. 8:4. A DOUBLE-WALLED SHELL
FIG. 8:5. BRACKET PRODUCED BY EXTRUDING PART AT LEFT AND SAWING IT INTO TWO PARTS

In some cases it is possible to combine two or more unsymmetrical parts into one symmetrical design that can be readily extruded. This proved to be practicable in the production of a number of brackets for aircraft. Typical is the bracket shown in Figure 8:5, which was produced as shown at the left and sawed in half to make two brackets as shown at the right.

Secondary Operations

Many of the parts produced by backward extrusion require only trimming to length. The surfaces are often sufficiently smooth to require no polishing, and tolerances can be held within reasonable limits in the process. A major advantage of impact extrusion here is the reduction in the machining required to finish a part. Certain operations that may be required on a part but cannot be performed during extrusion, include beading, flanging, flaring the open ends, drilling, and threading. Some examples of parts which have been through secondary operations are shown in Figure 8:6.

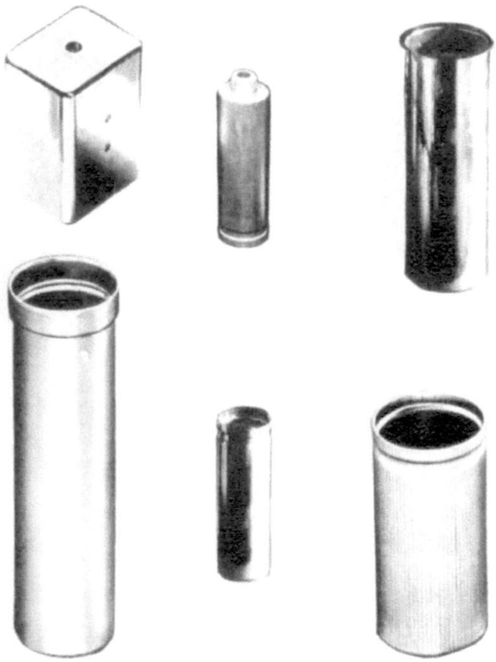

FIG. 8:6. SECONDARY OPERATIONS ARE USED TO PRODUCE FLANGES, FLARED ENDS, DRILLED AND THREADED PARTS

FORWARD EXTRUSION

In forward extrusion the punch is short and sturdy. It might be called a ram with a shaped head. It fits snugly into the lead-in portion of the die cavity to prevent backward extrusion of material. Therefore, there is no effect of a column to be considered, as there is in backward extrusion. In addition, the close fit prevents lateral motion of the punch during extrusion.

Unlike backward extrusion, where the length of the part is limited by the strength of the column of the punch, the length of an item produced by forward extrusion is limited only by the size of the blank that can be handled and the ease of removing the part from the die. Forward extruded parts can be made with length-to-diameter ratios of 100 to 1 or greater.

In backward extrusion, the shape of the blank has little effect on the shape of the part. On the other hand, in forward extrusion, the shape of the blank plays a role in shaping the

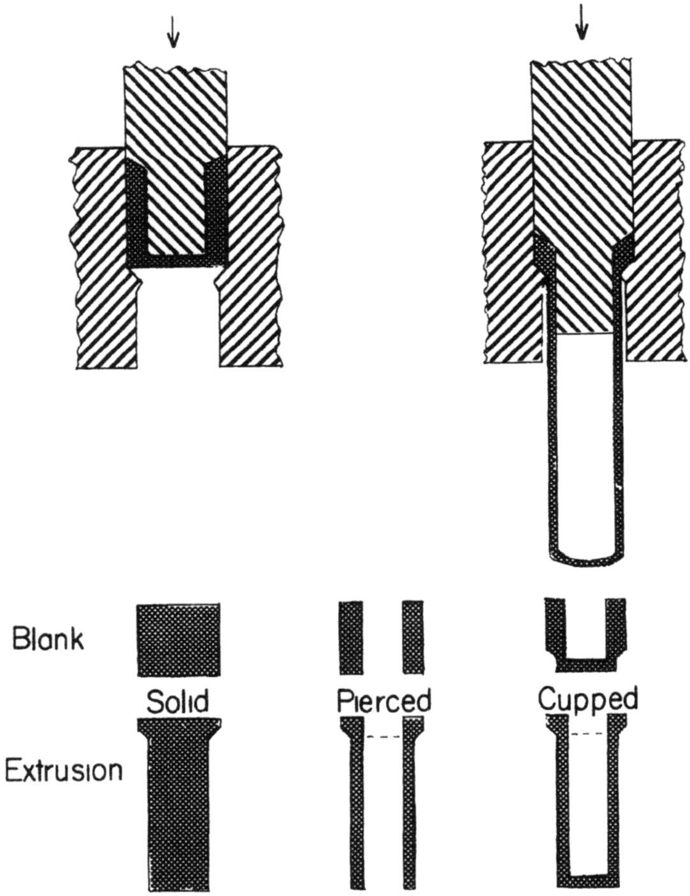

FIG. 8:7. A VARIETY OF PARTS PRODUCED BY CHANGING FORM OF BLANK

part. As shown in Figure 8:7, four different parts can be produced with one set of tools merely by changing the form of the blank. Of the two closed tubes shown, that at the top is more economical to produce because less machining of the blank is required. By changing the tip of the punch for one that tapers

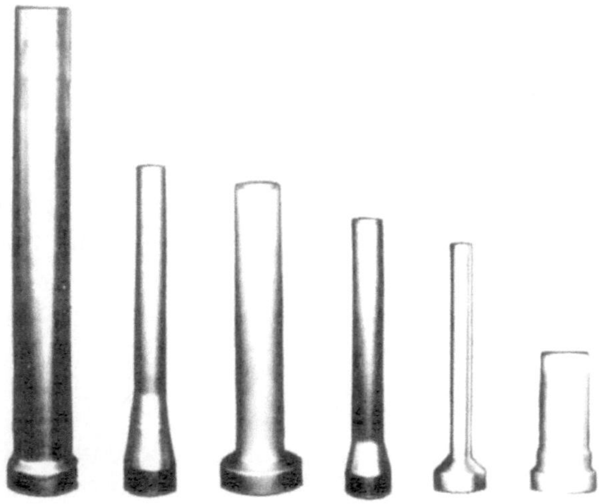

FIG. 8:8. PARTS PRODUCED BY FORWARD EXTRUSION

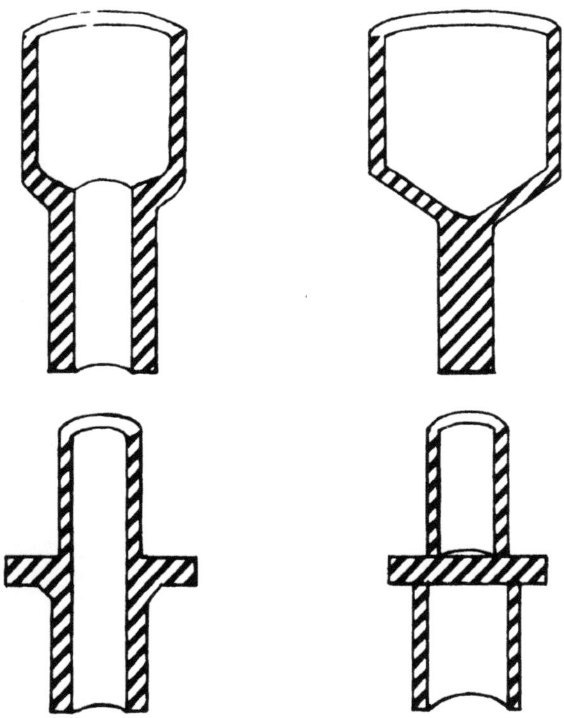

FIG. 8:9. PARTS PRODUCED BY BACKWARD AND FORWARD EXTRUSION

toward the orifice of the die, hollow parts can be made which have thicker walls at the bottom than at the top.

Forward extrusion cannot be used for so great a variety of parts as backward extrusion. In general, the parts best produced by forward extrusion are solid rods, or tubular forms such as those shown in Figure 8:8. The tubes can be open or closed, can have tapered walls, and the thickness of the tapered end can be changed. Flanges at the end can be heavy, stepped, or formed into various shapes. Ribs can be formed on some types of parts.

COMBINED EXTRUSION

Forward and backward extrusion can be combined to produce various parts. Most of the shapes discussed under backward extrusion can have forward-extruded rods or tubes formed

FIG. 8:10. STEPS IN PRODUCTION OF 75-mm SHELL
Top, left to right: Blank, Sized Slug, Backward Extrusion, Forward Extrusion; *Bottom*, left to right: Coining, Drawing, Nosing

at the same time. Sketches of typical parts are shown in Figure 8:9. The two sections of such a part can have the same or different cross-sectional areas, walls of different thicknesses, and the same or different shapes. For example, a cup having a square cross-section can be backward extruded at the same time as a tube, having a circular cross-section is being forward extruded. Tubes can be extruded in both directions from a flanged portion. Combined extrusion can often be used to produce a part that can take the place of an assembly. A one-piece tube with a cupped end is stronger than an assembly of a cup and a tube.

EXTRUSION AS A STEP IN PROCESSING

The use of cold extrusion as a step in a series of operations is quite common in the production of ferrous parts. Generally, the purpose is to eliminate several steps of deep drawing. Figure 8:10 shows steps in the production of a 75-mm steel shell. In the first step, the blank is preformed into a sized cup. In the second, the cup is backward-extruded. The third step is a forward extrusion to deepen the cavity. These steps are followed by coining, deep drawing, and nosing. It has been reported that the procedure outlined saves a number of steps in the production of the shell and reduces material requirements by 51 per cent.

Only a few examples of parts that can be produced by impact extrusion have been given in this chapter. Others will be discussed in the chapters dealing with specific metals. Many other types of parts could be shown but such figures would indicate only modifications and would serve no useful purpose. The producers of parts issue brochures containing sketches and photographs of various parts. In addition, numerous parts are illustrated in the following references:

REFERENCES

1. Barrett, J. J., Designing for Steel Extrusions, *Precision Metal Molding*, 17, May 1959:68.
2. Brown, D. I., Successful Design of Steel Parts for Cold Extrusion, Paper No. 22T34, American Society of Tool Engineers, meeting at Philadelphia, Pa, April 26-30, 1954.

3. EGAN, E. J., Jr. Aluminum Impacts Open New Areas for Design, *Iron Age*, 181, April 3, 1958:98.
4. FELDMANN, H. D., Properties and Possibilities of Using Cold Extruded Steel Parts, *Mem. Soc. Ingénieurs Civils de France*, July-August 1956:247.
5. FELDMANN, H. D., Design Principles for Cold Pressed Steel Components, *Konstruktion*, 3:82 (1959).
6. GABLER, P., Precision Impact Extrusion and Die Casting in Industry, *Aluminium*, 25:194 (1943).
7. VICTOR, M., Industrial Possibilities of Impact Extrusion, *Revue de l'Aluminium*, May 1957:535.

Chapter 9

ALUMINUM AND ALUMINUM ALLOYS

The major advances in the applications of impact extrusion have been in the fabrication of aluminum and its alloys. In roughly twenty years, applications have increased from the initial collapsible tubes and other containers to a wide range of parts for use in the aircraft, automotive, electronic, and other industries. Today a dozen or more impact-extruded forms are used in buses, trucks, and passenger cars in critical assemblies for braking, steering, and transmission. Aircraft use aluminum impact extrusions for the bodies of fuel pumps, bodies and cylinders for hydraulic landing gear, among a variety of applications. Other large forms are used in designs for advanced missiles. Condenser bodies, anodes, and other parts are used in electronic components. Industrial and domestic uses include bodies for grease guns and cases for flashlights.

MATERIALS

Aluminum containing no intentionally-added elements is available in various grades on a commercial basis. In this country, the grades range from 99 to 99.9% pure and in Europe, the range of purity extends to 99.99 per cent. This listing is not intended to imply that even higher degrees of purity cannot be obtained on special order, but the ranges quoted are commercial.

From the standpoint of volume, most impact-extruded parts are produced from commercial aluminum, nominally the 1100 grade which contains the minimum of 99% aluminum. When certain characteristics are required that are not obtainable with the 1100 grade, others are substituted. For example, as the

1100 grade is too stiff to form a satisfactory collapsible tube, a grade known as 1170, containing the minimum of 99.7% aluminum, is employed commercially in the United States for this service. On the other hand, the 1100 grade may not develop sufficient strength for a specific application and one of the alloys may be substituted.

It is probable that all of the wrought aluminum alloys can be impact-extruded and a number of them are in commercial use. Table 9:1 lists the composition of the most popular American alloys now being used in the production of impact-extruded parts.

TABLE 9:1

NOMINAL COMPOSITIONS OF SOME ALLOYS USED FOR IMPACT EXTRUSION

Alloy	*Composition %*						
	Al	Cu	Si	Mn	Mg	Cr	Zn
1170	99.7[a]	—	—	—	—	—	—
1100	99[a]	—	—	—	—	—	—
2014	rem	4.4	0.8	0.8	0.4	—	—
2024	rem	4.5	—	0.6	1.5	—	—
3003	rem	—	—	1.2	—	—	—
6053	rem	—	0.7	—	1.3	0.25	—
6061	rem	0.25	0.6	—	1.0	0.25	—
6063	rem	—	0.4	—	0.7	—	—
6151	rem	—	1.0	—	0.6	0.25	—
7075	rem	1.6	—	—	2.5	0.3	5.6

[a] Minimum

The manganese alloy, 3003, is frequently selected if moderately higher strength than can be obtained with the 1100 grade is desired. For somewhat higher strengths and greater machinability, magnesium-silicon alloys such as 6061 or 6151 can be used. For maximum strength, alloys of the copper group, such as 2014, or the zinc group, such as 7075, can be used. The latter develops the highest strength that has been attained in an impact-extruded aluminum part.

It should be noted that impact extrusion can also be used for the production of parts from clad materials. The mechanical properties are provided by the core. The cladding provides cathodic protection, as in other aluminum-clad alloys, and facilitates finishing. Parts may also be clad to simplify joining operations.

Nominal mechanical properties of a number of the alloys, in the annealed condition, are given in Table 9:2. As discussed in Chapter 3, relative ease of extrusion is indicated by strength and hardness. This table indicates that the 1170 grade is the softest and easiest to extrude and it will require the lowest extrusion pressure for a given part; whereas 7075 is the hardest to extrude and will require the highest extrusion pressure for the same part. Alloys such as 2014, 2024, and 7075 are often preheated several hundred degrees to reduce the pressure requirements.

TABLE 9:2

PROPERTIES OF ANNEALED ALLOYS
(In Order of Increasing Hardness)

Alloy	*Yield Strength*[a] *(0.2% offset)* *psi*	*Tensile Strength* *psi*	*Elongation (in 2 in.)* *%*	*Shear Strength* *psi*	*Brinell Hardness*
1170	—	10,000	50	8,000	19
1100	5,000	13,000	45	9,000	23
6063	7,000	13,000	—	10,000	25
6053	8,000	16,000	35	11,000	26
3003	6,000	16,000	40	11,000	28
6061	8,000	18,000	30	12,000	30
2014	14,000	27,000	18	18,000	45
2024	11,000	27,000	22	18,000	47
7075	15,000	33,000	16	22,000	60

[a] Compressive yield strength is approximately equal to the tensile yield strength

Approximate extrusion pressures required to produce a round shell by backward extrusion from various aluminum alloys are given in Table 9:3. These data were calculated on the basis of relative yield strengths and are merely indicative of the ranges to be expected. Actual pressures determined during the extrusion of a number of European alloys are given in Table 9:4, based on the work of von Zeerleder and von Berg.[1]

The American materials under discussion can be divided into two groups. The 1170, 1100, and 3003 grades can be hardened only by cold-working. The 1100 and 3003 grades have the advantage of minimum cost and therefore are selected in preference to the other materials if their properties permit. The remaining alloys listed in Table 9:1 can be heat-treated to improve the strength. These alloys are age-hardenable and are usually given a solution heat treatment followed by artificial aging (the -T6 condition) after impact extrusion to develop maximum properties. Unfortunately, no data appear to have

TABLE 9:3
APPROXIMATE EXTRUSION LOAD FOR ANNEALED ALLOYS
(Backward Extrusion)

Alloy	*Outside Diameter inches*	*Extrusion Load (tons) for Minimum Wall Thickness of*		
		0.015 in.	0.060 in.	0.125 in.
1100	1	51	27	18
	2	200	102	66
2014	1	143	76	50
	2	560	286	185
2024	1	112	59	40
	2	440	214	145
3003	1	61	32	22
	2	240	122	79
6053	1	97	52	35
	2	382	195	126
6061	1	97	52	35
	2	382	195	126
6063	1	71	38	25
	2	280	143	92
7075	1	153	81	54
	2	600	306	198

been published on the properties of the materials after impact extrusion, but the properties that can be expected are given in Table 9:5. This table indicates the strengths obtainable in the work-hardenable materials as well as properties developed by heat treatment of the age-hardenable alloys.

As applied to aluminum and its alloys, impact extrusion is a cold-working operation. Therefore, the process increases the hardness and strength of the materials. Table 9:6 shows the

TABLE 9:4
RELATIVE EXTRUSION PRESSURES FOR ALUMINUM AND SEVERAL ALUMINUM ALLOYS*

Composition %					*Brinell Hardness*	*Extrusion Pressure psi*
Al	Mn	Si	Mg	Cu		
99.99	—	—	—	—	14	55,000
99.5	—	—	—	—	23	74,000
rem	1.2	—	—	—	32	88,000
rem	0.2/1.0	0.5/1.5	0.5/1.0	—	34	98,000
rem	1.4	—	2.2	—	58	170,000
rem	0.5	—	0.55	3.8	58	183,000
rem	0.5	—	7.0	—	80	253,000

* Von Zeerleder and von Berg[1]

effect of backward extrusion on the yield strength. The softest materials are strengthened proportionally to a greater extent than the hard alloys. For example, the yield strength of the 1100 grade is increased roughly four times, whereas that of the 2014 grade is increased about three times. Heat treatment of the 2014 grade however, increases the yield strength considerably as is also shown in this table; the result with the 7075 grade is similar.

TABLE 9:5
TYPICAL PROPERTIES OF SOME WROUGHT ALUMINUM ALLOYS

Alloy and Temper	*Tensile Strength, psi*	*Yield Strength (0.2% offset) psi*	*Brinell Hardness*
1100-O[a]	13,000	5,000	23
1100-H14	18,000	17,000	32
1100-H18	24,000	22,000	44
3003-O	16,000	6,000	28
3003-H14	22-000	21,000	40
3003-H18	29,000	27,000	55
6061-O	18,000	8,000	30
6061-T6	47,000	42,000	95
6151-T6	48,000	43,000	100
2014-O	27,000	14,000	45
2014-T6	70,000	60,000	135
7075-O	32,000	14,000	—
7075-T6	76,000	67,000	—

[a] O, annealed temper of wrought materials
H18, full hard commercial temper
H14, material cold-worked to tensile strength approximately midway between fully annealed temper (−0) and full hard temper (−H18)
T6, solution heat treatment followed by artificial aging

The improvement in tensile strength resulting from backward extrusion with a reduction of 60 per cent is indicated in Table 9:7. These data were reported in *Alluminio*[2] and the compositions of the alloys were not given. The strengths of the softer alloys were roughly doubled by this reduction and even the hard alloys were improved considerably.

In an extensive investigation of backward extrusion, Fukui, Kudo, and Seino[3] included a determination of the improvement in hardness in aluminum and duralumin resulting from the

TABLE 9:6

IMPROVEMENT IN YIELD STRENGTH AFTER IMPACT EXTRUSION

Alloy	*Yield Strength psi*		
	Annealed	Impact-Extruded	Impact-Extruded Heat Treated
1100	5,000	19,000	—
3003	6,000	30,000	—
2014	14,000	39,000	55,000
6061	8,000	31,000	—
7075	15,000	41.000	70,000

TABLE 9:7

IMPROVEMENT IN TENSILE STRENGTH OF ALUMINUM ALLOYS REDUCED 60 PER CENT BY BACKWARD EXTRUSION*

Alloy[a]	*Extrusion Pressure, psi*	*Tensile Strength, psi*	
		Annealed	Impact-Extruded
Al 99	63,000	13,000	28,400
Al-Mg	71,000	15,500	33,000
Al-Si-Mg	100,000	18,500	34,000
Al-Cu-Mg	120,000	27,000	44,000
Al-Zn-Mg-Cu	135,000	31,000	45,500

[a] Composition not given
* *Alluminio*[2]

process. They worked with soft and hard tempers of aluminum and decided that the backward extrusion of soft material resulted in a considerable increase in hardness, whereas there was little change in hardness from the extrusion of hard material. They also concluded that speed of extrusion had no effect on the properties developed. Their data are given in Table 9:8.

TABLE 9:8

IMPROVEMENT IN HARDNESS OF ALUMINUM ALLOYS REDUCED 80 PER CENT BY BACKWARD EXTRUSION*

Material	*Condition*	*Initial Vickers Hardness*	*Final Vickers Hardness*		
			Upper Wall	Lower Wall	Bottom
Aluminum	Annealed 750 F, 1 hr	20	44	44	47
Aluminum	Hot-extruded and cold-upset	43	51	50	50
Duralumin	Annealed 790 F, 1 hr	63	124	131	140

* Fukui, Kudo, and Seino[3]

SPECIFICATIONS

There appears to be only one specification of national scope covering impact-extruded parts produced from aluminum and its alloys. This is military specification MIL-A-12545 (ORD). It covers commercial aluminum, grade 1100, and the alloys 2014, 6061, and 7075. No mechanical properties are specified for the materials in the extruded condition but such properties are specified for annealed and heat-treated parts produced from the 2014, 6061, and 7075 alloys. Tests are to be made on specimens cut from the side walls of the impact-extruded parts. The specifications also require hardness to be determined on the bottom of any part that is too small to permit tension specimens to be cut from the side wall of the extruded piece. Requirements for tension and hardness are given in Table 9:9 and comparison with Federal Specification QQ-A-367 (Aluminum Alloy Forgings, Heat Treated) shows that the two sets of requirements are practically identical.

TABLE 9:9

REQUIREMENTS OF IMPACT-EXTRUDED PARTS FROM ALUMINUM ALLOY MIL-A-12545 (ORD)

Alloy and Temper	*Tensile Strength psi*	*Yield Strength (0.2% offset) psi, min*	*Extension Under Load in./in., min*	*Elongation (4D) min %*	*Brinell Hardness min*
1100-F	—	—	—	—	—
2014-F	—	—	—	—	—
2014-O	30,000 max	—	—	—	—
2014-T4	55,000 min	32,000	0.0050	10	100
2014-T6	65,000 min	55,000	0.0072	6	125
6061-F	—	—	—	—	—
6061-O	22,000 max	—	—	—	—
6061-T4	26,000 min	16,000	0.0036	12	50
6061-T6	38,000 min	35,000	0.0055	6	80
6061-T84[a]	35,000 min	30,000	0.0050	4	75
[b]	38,000 min	35,000	0.0055	4	75
7075-F	—	—	—	—	—
7075-O	40,000 max	—	—	—	—
7075-T6	75,000 min	65,000	0.0083	5	135

[a] Wall 0.040 to 0.093 inch thick
[b] Wall 0.094 inch thick or more
KEY: F as fabricated
O annealed
T4 solution heat-treatment
T8 solution heat-treatment, artificially aged
T86 solution heat-treatment, cold-worked, artificially aged

SIZES AND TOLERANCES ON PARTS IN COMMERCIAL PRODUCTION

The cross-sectional areas and lengths of parts that can be produced by impact extrusion depend on the equipment available but they are also affected by limitations in the process itself. In backward extrusion, the strength of the column of the punch is a limiting factor. For 1100 and similar materials, the limit of length is generally eight to ten times the diameter of the shell; for the stronger alloys, it may be five times. Parts up to 14 inches in diameter and 60 inches long have been produced and parts up to six inches in diameter and 20 inches long are in regular production. As long as the ratio between diameter and length is maintained, the size of the part is limited only by the capacity of the press. Figure 9:1 indicates the range of sizes being produced.

FIG. 9:1. RANGE OF SIZES PRODUCED BY IMPACT EXTRUSION
Hand, holding typical small shell is dwarfed by the largest part made to date, a cylinder 8.5 inches in diameter and 34 inches long (*courtesy* Aluminum Company of America)

An indication of the thicknesses of walls and the diameters available in regular production is given in Table 9:10. As the hardness of the alloy increases, there is a corresponding increase in the minimum thickness of wall that can be produced on a competitive basis. This does not mean that it is impossible to produce thinner walls than those listed but the economics of

TABLE 9:10
RANGE OF COMMERCIAL SIZES OF BACKWARD EXTRUDED PARTS

Alloy	*Wall-Thickness in.*		*Diameter in.*	
	Max	Min	Max	Min
1070	0.010	0.0035	2.5	$\frac{3}{8}$
1100, 3003	0.5[a]	0.010	6	$\frac{3}{4}$
6061, 6151	0.5[a]	0.040	5	$\frac{3}{4}$
2014, 7075	0.5[a]	0.062	4	$\frac{3}{4}$

[a] Nominal; maximum wall-thickness not established

the operation may make it impractical because of high costs of tooling.

In forward extrusion, the punch can be considerably more sturdy than that used in backward extrusion and strength of column is not a limiting factor. In forward extrusion, the ends of tubes may be open or closed, and ends may be of any reasonable thickness from 0.020 inch to several inches. Tubes up to six inches in diameter and 100 inches or longer are in regular production and increase in press capacity may soon raise these limits to at least 12 inches in diameter. Walls range from 0.015 to 0.25 inch or more in thickness.

One of the major advantages of impact extrusion is the ability to hold close dimensional tolerances on the thickness of wall and bottom, and both outside and inside diameters. The length shows greater variation and the tolerance that can be held commercially depends on a secondary operation—trimming the shell. An indication of tolerances that can be expected on parts made from aluminum alloys is given in Table 9:11 which is based on figures published by a number of producers. These tolerances are sufficiently close in most cases to obviate the necessity of a sizing operation. In addition if the tools are maintained well polished, the walls of the shell have a finish that is often satisfactory with no additional finishing operations. Surface finishes obtainable on the part as extruded, range from 40 to 250 microinches *rms*, depending on the shape of the part and the alloy.

Production rates on aluminum alloys are relatively high. Collapsible tubes and simple can-shaped parts are produced at rates as high as 40 to 80 parts a minute. Rates for engineering components, produced from the stronger alloys, vary greatly.

One company reports that its highest production rate is 45 parts an hour; others state that rates up to 250 to 500 parts an hour can be maintained without difficulty.

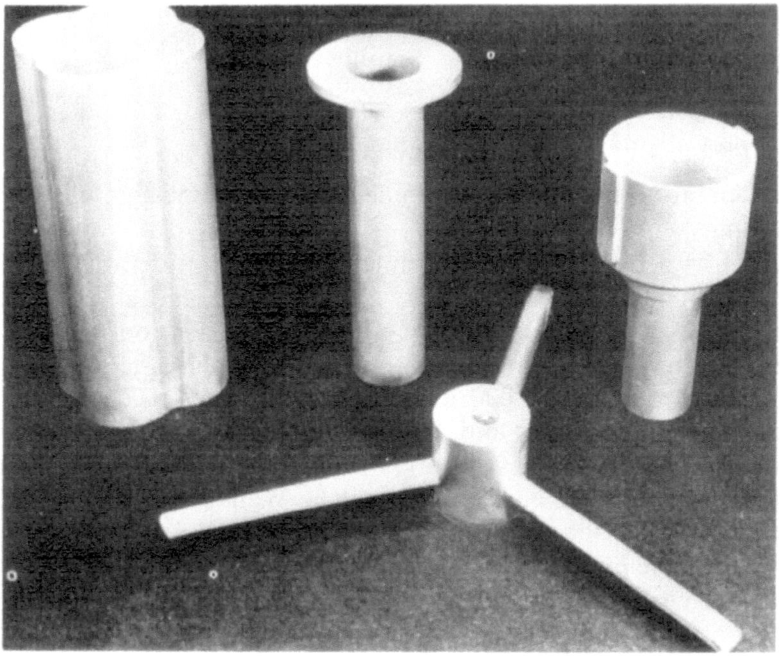

FIG. 9:2. LATERAL IMPACT EXTRUSION: NEW TECHNIQUE DEVELOPED BY ALCOA.
A lateral extrusion is shown surrounded by a backward extrusion (*left*); a forward extrusion (*center*); and a combination backward-and-forward extrusion (*right*)
(*courtesy* Aluminum Company of America)

APPLICATIONS

As its advantages become more widely recognized, impact extrusion is being extended to more and more fields. The selection of this process, of course, hinges on its usefulness for the production of a specific part, as has been discussed in a preceding chapter. When the shape of the part permits use of impact extrusion, either as a direct substitution for another manufacturing method or with a change in design, the process often permits considerable savings in material and machining.

Some examples of commercial parts are given in Figures 9:2 and 9:3.

Although many parts are produced with no additional operations except trimming to length, others require secondary operations, such as spinning a neck, bulging, piercing, drilling, or threading. The close tolerances and excellent finishes achieved by impact extrusion, however, reduce secondary operations to the minimum.

TABLE 9:11

TYPICAL TOLERANCES FOR REGULAR SHAPES IN ALUMINUM ALLOYS*

Outside Dimensions in.	*Wall Thickness in.*	*Bottom Thickness in.*	*Tolerances* O.D. ±in.	I.D. ±in.	Bottom ±in.	Length ±in.
Round Shells						
$\frac{1}{4}$	0.010-0.020	0.020-0.030	0.005	0.006	0.007	$\frac{1}{32}$
1	0.015-0.025	0.025-0.035	0.005	0.006	0.010	$\frac{1}{32}$
2	0.020-0.030	0.032-0.042	0.007	0.008	0.012	$\frac{1}{32}$
3	0.025-0.050	0.040-0.065	0.008	0.010	0.015	$\frac{1}{32}$
Square Shells			Side	Side		
$1\frac{3}{4} + 1\frac{3}{4}$	0.025	0.035	0.008	0.010	0.015	$\frac{1}{32}$
2×2	0.025	0.035	0.008	0.010	0.015	$\frac{1}{32}$
$2\frac{1}{4} \times 2\frac{1}{4}$	0.062	0.075	0.008	0.010	0.015	$\frac{1}{32}$
Rectangular Shells			Side	Side		
2×1	0.025	0.035	0.008	0.010	0.015	$\frac{1}{32}$
$2\frac{1}{2} \times 2\frac{1}{4}$	0.031	0.050	0.008	0.010	0.015	$\frac{1}{32}$

* SOURCE: Various producers

To furnish an idea of the variation in the applications of impact extrusion, some specific examples are discussed briefly in the following sections.

In Aircraft

The aircraft industry was among the first to recognize the advantages that could be obtained through the impact extrusion of aluminum parts, and the experience has been carried forward to the production of parts in other aerospace components. A few examples will indicate the many applications that have been made.

A plunger for landing-gear, in the form of a flanged tube, has been forward-extruded from 2014 alloy and subsequently heat-treated to the -T6 condition.[4] The length of the plunger is

21 inches over-all, and the inside bore was 19 inches deep. The outside diameter was 1.750 ± 0.010 inch and the inside diameter, 1.500 ± 0.010 inch. Surface roughness was 63 microinches,

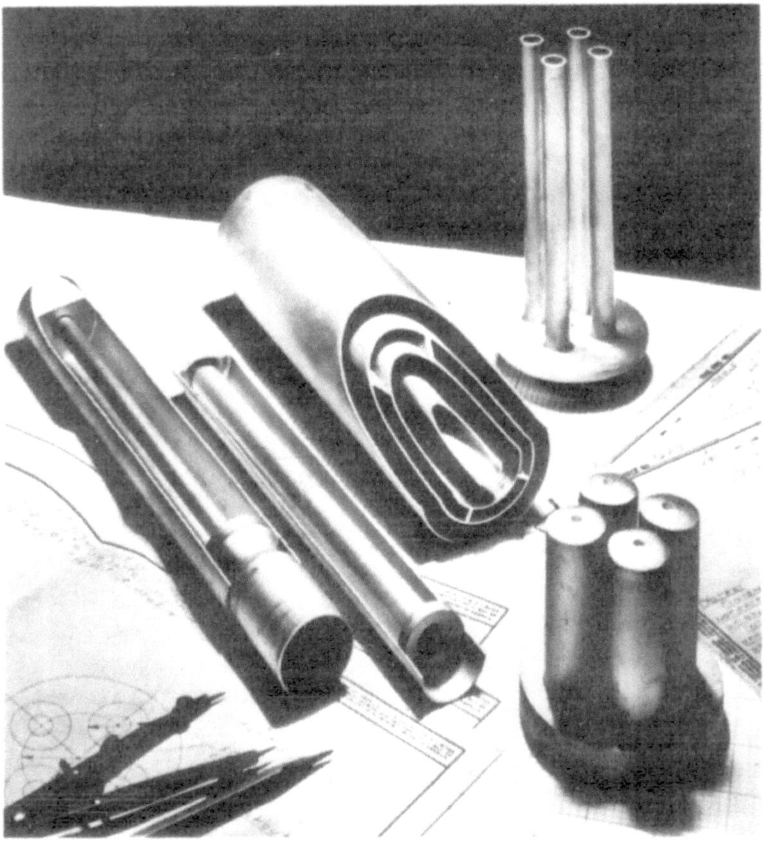

FIG. 9:3. COMPLEX EXTRUDED COMPONENTS OF ALUMINUM FOR NUCLEAR APPLICATIONS
At center is a two-piece assembly of inner double-walled part with longitudinal ribs seated within a second double-walled part (*courtesy* Aluminum Company of America)

rms. To complete the part, machining was required only on the closed end and on the flange at the open end.

It is sometimes advantageous to combine two or more parts to obtain a simple shape for extrusion rather than to attempt to extrude the single component. Quadt[4] notes that this procedure has been used to produce attachment fittings for airframes from

2014 alloy. A rectangular can with walls 0.130 inch thick was backward-extruded from the alloy and cut in half lengthwise to form two fittings. The parts were subsequently heat-treated to the -T6 condition.

A rib for the nose of an aileron was produced by three processes for a comparison of costs. The component was produced from 2014 alloy by impact extrusion, and from 7075 by machining, both from a billet and by forging, followed by machining for good finish. The investigation showed that in production quantities of 1000, the impact extrusion cost only 35 per cent as much as the part machined from a billet; and only 67 per cent as much as a part that had been forged and machined.

A solid impact-extruded part has been used as a blank for the production of forged compressor blades for jet engines. Producing the preform by impact extrusion resulted in considerable savings in both material and processing.

Other aircraft applications include fittings for bomb-bays, hydraulic pistons, a housing for the float in an aircraft fuel gage, and a component for the afterburner of a jet engine.

Automotive Components

In the automotive field, applications of impact extrusion include torque reaction shafts and servo pistons for transmissions, oil filter cases, pistons for hydraulic brakes and reservoirs for power brakes, baffle tubes for shock absorbers and covers for spark plugs. Some decorative items are also impact-extruded including elements for a radiator grill.

Aluminum pistons for automotive service have been backward-extruded from some of the stronger aluminum alloys.[5] The alloy blank is preheated before being fed to the press and the operation is completed in one stroke. After it is trimmed to length, the piston is heat-treated to develop the required mechanical properties, and finished by conventional machining procedures. The impact-extruded piston has thinner walls than the conventional piston produced by a permanent mold casting. In one design, the extruded piston was almost two ounces lighter than the corresponding cast piston. The wrought structure resulting from extrusion is stronger than the cast structure, a factor which shows up in tests for fatigue. No cracks from fatigue appeared in an impact-extruded piston after 40 million cycles of stress under the same conditions that had caused failure of a cast piston in 10 million cycles.

Electrical and Electronic Devices

Among electrical parts produced by impact extrusion are frames for a variety of electric motors; a housing for an electric motor with a closed end, designed to hold the bearing and locate the field winding; and an electrical cable connector having grooves on all sections of the inner surface, to ensure proper alignment during installation. Other items include a housing for a lightning arrester, used to protect a telephone circuit; and various structural components for radar.

A rotor housing was originally fabricated by brazing a machined head to a length of thin-walled tubing, the construction requiring 17 cubic inches of aluminum.[6] The part is now produced by impact extrusion with one stroke of the press, using a blank that has a volume of 6½ cubic inches. As a result, the material required was reduced 60 per cent, and there was a reduction of seven minutes per unit in machining and assembly time.

In the production of complex parts, blanks or tools may be heated moderately to ease the extrusion. This procedure was used for the production of several electronic components, including a trimmer condenser and an anode for an electrolytic condenser.[7]

The trimmer condenser has been produced at the rate of 80,000 components per week. The stator is backward-extruded with three concentric sleeves, the rotor with four. The wall of each sleeve is 0.014 inch thick and 0.26 inch deep; the space between sleeves is 0.035 inch. A collar, 0.095 inch long by 0.204 inch in diameter, is forward-extruded on the rotor. Three bosses, 0.060 inch in diameter, are forward-extruded in the end faces of both stator and rotor to serve as integral rivets for assembling the condenser. Tubular blanks of 99.5% aluminum are used for the parts. They are preheated in grade 40 lubricating oil to 200 F before being fed to the extrusion press. A production rate of 750 parts an hour is maintained in fabricating these parts.

The anode for the electrolytic condenser is extruded in the same manner from a preheated solid blank of 99.5% aluminum.

This component is $2\frac{5}{8}$ inches long by 0.93 inch in diameter and has six concentric sleeves. The wall of the sleeves is 0.145 inch thick and the spacing between sleeves is 0.051 inch. A recess, 0.145 inch in diameter by 0.200 inch deep, in the small end of the anode, is formed by a second backward extrusion. The component is trimmed to length and the sides are slotted in a subsequent operation.

General Industry

In the industrial field, aluminum impact extrusions are produced in lengths up to five feet for use as accumulators, pistons, cylinders, and fluid-power components. Other applications include oil filters and housings for pneumatic tools. An air cleaner and dryer shell for use on the compressed-air line of a paint sprayer is another example.

A shell $2\frac{1}{2}$ inches in diameter by 12 inches long, for a dry-type fire extinguisher, is impact-extruded from grade 1100 aluminum. This part took the place of a deep-drawn shell at a considerable reduction in cost.

A ribbed cylinder for a grease gun is produced by impact extrusion, the ribs serving to strengthen the part and prevent slippage in use. This one extruded piece took the place of an end-plug and tube, at a considerable saving in costs for assembling.

An air filter, originally made by welding a steel end-plate to a steel tube was redesigned in one piece and made from aluminum by impact extrusion. Costs of welding were eliminated, and supporting brackets were not required because of the reduction in weight resulting from the substitution of steel by aluminum.

A diaphragm bellows chamber for a pressure regulator was originally machined from bar stock. Conversion to backward extrusion resulted in saving over 35 per cent in materials and labor.[8]

Gabler[9] points out that a flanged sleeve can be produced by backward extrusion, using a hollow punch and a projection on the base of the die as the extrusion edges. A ring-shaped blank is used and the material is extruded from the inner edge of the ring. One application of this procedure is an *industrial spool* produced from commercial aluminum. The second flange is formed later by spinning. Changing from deep drawing to impact extrusion eliminated five steps in the production of these spools.

Ordnance Components

Among ordnance components produced from aluminum or aluminum alloys by impact extrusion are tetryl cups for artillery shell, casings for depth charges, tail fins and bodies for projectiles, and a number of items for guided missiles. A rifle grenade tube is produced by a combination of backward and forward extrusion leaving a web between the sections; a rifle telescope tube is produced from 6061 alloy and heat-treated to the -T6 condition.

Quadt[10] reports that a motor five inches in diameter, for the Sidewinder missile, is extruded in one piece with a ribbed channel integral with the tube of the motor.

A propellant capsule, 9 inches long by $2\frac{1}{2}$ inches outside diameter, had been produced by welding a forged end to a length of extruded tubing.[6] The method required 14 cubic inches of aluminum and production time was $13\frac{1}{4}$ minutes for each unit. Converted to backward extrusion, the part was produced from a blank having a volume of $6\frac{2}{3}$ cubic inches and production time, including trimming and finishing, was only five minutes.

In many bars and tubular shapes of various diameters, impact extrusion can be used to advantage, sometimes with considerable savings in material. Quadt[11] has discussed a sleeve for a fuse, composed of 2014-T6, that was originally produced from tubing on an automatic screw machine. Over 75 per cent of the material was lost in the machining operation. Produced by impact extrusion, with tolerances of 0.003 inch, the part cost considerably less than the unmachined tube.

One double-walled ordnance component produced by impact extrusion from aluminum is $3\frac{3}{8}$ inches long by $2\frac{3}{4}$ inches in diameter and has an internal tube 0.65 inch in diameter. The piece requires only trimming to length because the surface finish and tolerances resulting from the operation are satisfactory for the functioning of the part.

Containers

From the point of view of quantity of production and length of time the application has been employed, the collapsible tube is the most widely used aluminum container. This subject will be discussed in detail in Chapter 10 in connection with the applications of lead and tin and therefore it suffices to note that the aluminum collapsible tube is produced from the 1170 grade and is cold-worked during extrusion. The tube must therefore be annealed to make it soft enough for service.

Gotch, Eike, and Brighton[12] have reviewed the status of the use of aluminum for food cans and report that impact-extruded cans are used extensively in Europe. These containers are used for beer, pressurized products, powdered coffee, and specialty foods. They are also used to package pharmaceutical preparations and horticultural products. Many products for photography, including color films, are packaged in aluminum impact-extruded cans. Intricate engineering components are frequently packaged similarly to protect them during shipment.

In changing from the common tin can to aluminum, the exact increase in gage of the material must be determined for each size of can and each use. To prevent denting and other damage from handling, it is probable that the gage must be increased 35 per cent over that used for the conventional tinned steel can.

For the protection of food products, the interior of aluminum cans should be lacquered, and the degree of protection will depend on the acidity of the product. Although the oxide film on aluminum is self-perpetuating in air, Gotch, Eike, and Brighton[12] point out that there is no oxygen available to repair the damaged surface of films in the airtight container, and tests have shown that the aluminum can is not so resistant to food products as is tin plate.

Aerosol cans are produced by backward extrusion of 1100 grade aluminum. The neck is spun in a subsequent operation. The wall of these cans is 0.014 inch thick. Cans of this type were produced originally as a three-piece assembly; impact extrusion permitted the combination of two of these three parts.[8]

A recent development in manufacturing containers is the combination of continuous casting and impact extrusion to produce a beer can.[13] Pigs of 1100 grade aluminum are melted in an induction furnace and continuously cast on a wheel that produces strip $\frac{1}{2}$ inch thick and 4 to 8 inches wide. The strip leaves the wheel at a speed of 10 to 15 feet a minute but it is expected that the speed can be increased greatly. The strip passes directly to a rolling mill which reduces the thickness to

$\frac{1}{8}$ inch and to a blanking press which produces as many as 3000 pounds of blanks an hour. Blanks are annealed, lubricated, and delivered to the extrusion press, which produces cans with a wall 0.013 inch (+0.000 to 0.001 inch thick) at the rate of 3600 an hour. The cans are about 60 per cent lighter than conventional beer cans of the same size, which results in a saving in weight of about $1\frac{1}{2}$ pounds for a case of 24 cans.[14] The cost of

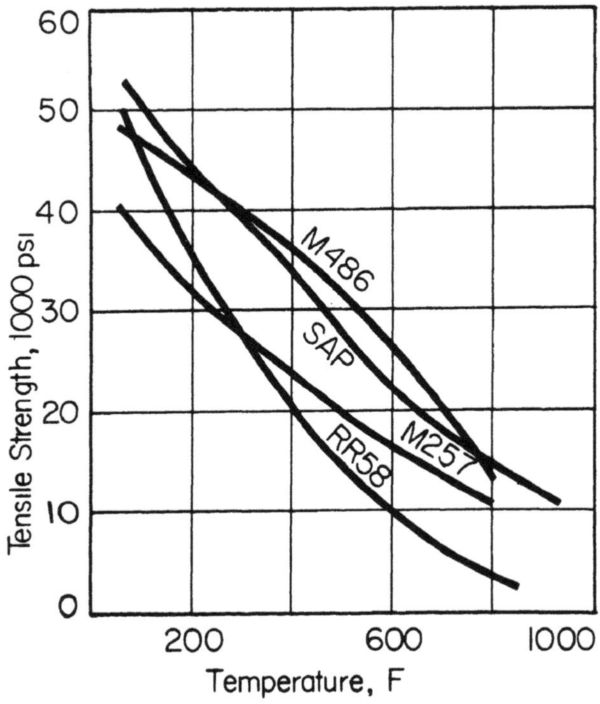

FIG. 9:4. COMPARISON OF TENSILE STRENGTHS AT HIGH TEMPERATURES OF PRODUCTS FROM SINTERED ALUMINUM AND FROM CONVENTIONAL ALUMINUM ALLOY (Doyle[16] and Towner [17])

the aluminum can is reported to be close to that of the conventional beer can (0.009 inch thick). This development is important because it permits the brewery to install an integrated can line for its own use. Similar intallations are being investigated for handling frozen concentrates of orange juice. A side-light is the expectation that littering of roadsides with discarded cans will be reduced because of the salvage value of the aluminum.

Housewares

For household use, vacuum bottles with decorative ribs, coffee makers, cocktail shakers, lamp housings, and vases are produced in a variety of attractive designs by impact extrusion. Items of hardware include door check cylinders and door latch housings.

The housing for a heating element in an automatic dishwasher was originally produced as a four-piece assembly that required four parts produced on a screw-machine. It was redesigned into an assembly requiring just two impact-extruded parts. The elimination of two components and the reduction in assembly time of the unit saved over 50 per cent of the cost of the housing.

A cylinder for a door closer, formerly made by assembling a tube and a screw machine part, required an O-ring seal to prevent leakage.[8] Redesign made it possible to convert this part to one piece by backward extrusion, the O-ring seal was eliminated, and there was no more leakage.

The barrels in many conventional door locks are composed of a deep-drawn flanged part held between two mating plates. Crimping one of the plates completes the assembly. A simple change in design permitted the production of components serving the same function from 1100 grade aluminum. Cold working resulting from the extrusion operation produced a part as strong as the original.

SINTERED ALUMINUM POWDER

A relatively recent development that holds promise of extending the temperature limits for aluminum alloys is the production of sintered aluminum powder alloys in Switzerland. The availability of this material was announced some years ago and its properties were discussed by Irmann.[15] Only within the past few years, however, has it become a commercial product, or rather, a series of products.

Briefly, the material is produced from aluminum flake, of the type usually employed in paint pigments, rather than from the granular powder of various shapes which is used in powder metallurgy. As each flake of the aluminum has a coating of oxide, it is possible by varying the thickness of the coating to vary the oxide content of the sintered aluminum powder, usually called SAP.

In the production of shapes from SAP, the powder is compacted, sintered, and then rolled, forged or extruded into sheet, rod, bar, or other shape. Alternately, the compacted mass can be hot-extruded into a rod. For the production of items by impact extrusion from SAP, blanks of suitable thickness are cut from the rod and either backward- or forward-extruded in the same manner as blanks of wrought material.

While investigating the properties of SAP, the Swiss discovered that the material retained its strength to higher temperatures than were useful for conventional aluminum alloys. The properties of SAP containing 9 to 11 per cent of aluminum oxide were compared with those of a wrought alloy that had been developed by Doyle[16] for service at elevated temperatures. The wrought alloy. RR 58, is used for forged components in gas turbine engines operating at 400 to 500 F, which is about the limit of useful service for conventional aluminum alloys. The effects of temperature on the tensile strength of SAP and RR 58, as determined by Doyle, are shown in Figure 9:4.

In the United States, the Aluminum Company of America has also been working with aluminum powder metallurgy products, which they have called APM alloys. Alcoa has developed several grades, containing varying quantities of aluminum oxide, and have made these alloys available commercially. They have also developed APM products based on an entirely different strengthening mechanism. For these products, atomized powder rather than flake is used, and the oxide content is relatively small. They are strengthened by adding alloying elements that form insoluble intermetallic compounds with the aluminum. Both types of material were discussed by Towner.[17] For comparison with the materials investigated by Doyle, the properties of APM M257 (5% Al_2O_3) and M486 (15% $FeAl_3$) are also plotted in Figure 9:4. It is apparent that all of these powder metallurgy products retain their strength to higher temperatures than does RR 58. As a matter of fact, these materials are approximately as strong at 800 F as RR 58 is at 500 F.

The mechanical properties that can be expected in impact-extruded parts from aluminum powder are indicated in Table 9:12; data were determined on samples cut from the side walls of the parts. A comparison of these data with those appearing in Tables 9:6 and 9:7 shows that the properties of the parts

TABLE 9:12
MECHANICAL PROPERTIES OF IMPACT-EXTRUDED PARTS FROM ALUMINUM POWDER

Material	*Tensile Strength psi*	*Yield Strength*	*Elongation (in 2 in.)*	*Brinell Hardness*	*References*
SAP (7% Al_2O_3)	35,000	30,000	11	65	10
SAP (12% Al_2O_3)	51,000	40,000	6.5	83	10
M257 (5% Al_2O_3)	40,000	28,000	17[a]	—	17
M486 (15% $FeAl_3$)	49,000	32,000	16[a]	—	17

[a] Elongation in four diameters

from powder compare favorably with those of " as-extruded " conventional alloys.

Applications

Applications for impact extrusions of these APM (aluminum powder metallurgy) products are just being developed. Examples of prototype parts are shown in Figure 9:5. Particular interest

FIG. 9:5. PROTOTYPE PARTS FROM APM BLANKS
Showing finned tube, one-quarter inch diameter, producible in lengths to fourteen feet from pierced slug shown at center (*courtesy* Aluminum Company of America)

has been expressed in the possibility of using them in aircraft and in nuclear reactors because they can withstand higher temperatures than are applicable for conventional aluminum alloys.

Actuating cylinder parts, $2\frac{3}{4}$ inches outside diameter by 2 inches inside diameter, have been impact-extruded from both M257 and M486 alloys.[17] For several years, this component has been produced from M257 by forging. Experimental hollow shells have been impact-extruded from M257 in sizes up to 6 inches in diameter with a 1-inch wall and weighing up to 40 pounds.

Finned tubing, having up to 10 fins, has been forward-extruded in lengths to 5 feet, from M257. This tubing is 0.305 inch inside diameter with a 0.030 inch wall; the outside diameter of the fins is 0.51 inch. The product is being developed for use in nuclear reactors cooled with organic liquids.

Siegel and Wilson[18] have suggested the use of M257 as a structural element to support the fuel in a nuclear reactor that employs UO_2 as the fuel. They report that the part produced from powder can be used up to 800 F, which is 300 degrees higher than the useful limit for other structural aluminum alloys.

REFERENCES

1. Von Zeerleder, A., and von Berg, K., Relations of the Power Demands in Extrusion of Pure Aluminum and Various Aluminum Alloys, *Schweizer Archiv*, 3:109 (1937).
2. Impact Extrusion of Aluminum, *Alluminio*, 28:75 (1959).
3. Fukui, S., Kudo, H. and Seino, J., Study on Impact Extrusion Method for Nonferrous Metals, *Reports, Institute Science and Technology, University, Tokyo*, 11:91 (1957).
4. Quadt, R. A., Designing Cold Impact Forgings, *Machine Design*, October 1952:190.
5. Extruded Pistons, *Modern Metals*, July 1954:82.
6. *Impact Extrusion*, Pheoll Manufacturing Company.
7. Impact Extrusion, *Machinery* (London), July 3, 1947:3.
8. *Coldprest Impact Extrusions*, Mueller Brass Company.
9. Gabler, P., Precision Impact Extrusion and Die Casting in Industry, *Aluminium*, 25:194 (1943).
10. Quadt, R. A., Latest Developments in Impact Extruded Aluminum Alloys, SAE paper 65D, June 15, 1960.
11. Quadt, R. A., When to Use Cold Extrusions of Aluminum, *Design Engineering*, November 1956:31.

12. GOTCH, L. P., EIKE, E. F., and BRIGHTON, K. W., The Status of Aluminum for Food Cans,' *Light Metal Age*, April 1959:8.
13. Continuous Casting, Impact Extrusion Spark Revolution in Aluminum Part Making, *Steel*, January 12, 1959:7.
14. Aluminum Cans, *Light Metal Age*, August, 1958:10.
15. IRMANN, R., Sintered Aluminum with High Strength at Elevated Temperatures, *Metallurgia*, September 1952:125.
16. DOYLE, W. M., The Properties of Hiduminium 100 (SAP) in Sheet Form, *Sheet Metal Industries*, 32:889 (1955).
17. TOWNER, R. J., *Alcoa's APM Alloys*, Aluminum Company of America, September 15, 1960.
18. SIEGEL, S., and WILSON, R. F., OMCR Power Plants; Their Status and Promise, *Nucleonics*, 17, November 1959:118.
19. *Alcoa Impacts—Metal in Motion*, Aluminum Company of America.

Chapter 10

LEAD, TIN, COLLAPSIBLE TUBES

Lead, tin, and their dilute alloys differ from the other materials discussed in this book in one particular; they recrystallize at nearly room temperature. Consequently, the rise in temperature resulting from impact extrusion is sufficient to exceed their temperatures of recrystallization and the alloys undergo hot working during extrusion. As a result, the lead and tin alloys currently being impact-extruded are soft and ductile as they come from the press and no annealing is required.

LEAD AND LEAD ALLOYS

Although all the common lead alloys can be impact-extruded, industrial applications have generally been limited to common desilverized lead and the antimonial lead alloys. To some of the latter, tin or silver is added. This subject will be discussed under collapsible tubes. The mechanical properties that can be expected in impact-extruded lead parts are indicated in Table 10:1.

A disadvantage of common lead is the low strength that can be obtained; therefore antimonial lead is generally used to produce stronger parts. About 3.5 per cent of antimony is soluble in lead at 477 F, and the solubility falls to about 0.25 per cent at room temperature. This change in solubility indicates that antimonial lead can be age-hardened. Although the property is well known, no application seems to have been made to impact-extruded parts. As shown in Table 10:1, antimonial lead is considerably stronger than common lead in the annealed condition, and the difference can be accentuated markedly by a simple heat-treatment.

TABLE 10:1
STRENGTHS OF LEAD AND TIN ALLOYS USED FOR IMPACT EXTRUSION

Composition, %				Tensile Strength, psi
Pb	Sn	Sb	Cu	
99.7	—	—	—	2500
99	—	1	—	3000
96	—	4	—	4000[a]
96	—	4	—	12,000[b]
—	99.8	—	—	2200
—	99.6	—	0.4	3300
—	99	—	1	7100

[a] Cold rolled 95%
[b] Heated to 450 F, water-quenched, aged one day at room temperature

The major application of impact extrusion of common lead and antimonial lead is in the production of collapsible tubes. Closures for the tubes are also impact-extruded but, in many cases, plastic caps have displaced lead caps.

The lead-tin alloys (solders) can be readily impact-extruded also and this procedure appears to offer a simple method of preparing thin-walled tubular preforms for soldering operations. Apparently, no application of this type has been made.

Some of the high-lead-bearing metals have been impact-extruded experimentally. No commercial application seems to exist, although this procedure would be a simple way to prepare shouldered bushings.

TIN AND TIN ALLOYS

Straits tin (99.8% Sn) has been used to prepare impact-extruded parts. Generally, however, tin hardened with copper is preferred. As shown in Table 10:1, the tin-copper alloy is considerably stronger than Straits tin. The major application of these alloys is the production of collapsible tubes which will be discussed in the following section.

It should be possible to impact-extrude the high-tin solders as readily as the high-lead varieties and probably it is also possible to extrude some of the tin-base-bearing metals. No experimental work has been reported on these alloys, however, and there are no commercial applications at present.

COLLAPSIBLE TUBES

The collapsible tube has been defined as

> . . . a cylinder of pliable metal which can be sealed in such a manner that its contents, although readily discharged in any desired quantity, are protected from contact with air or moisture during the whole period of use as a container.[1]

Probably the most familiar example is the tube for toothpaste. Although various methods have been developed to produce collapsible tubes from foil, the only commercially satisfactory method appears to be impact extrusion.

Materials

Collapsible tubes are produced from lead, antimonial lead (sometimes containing tin or silver), various proprietary lead alloys, tin-coated lead alloys, tin, tin-copper alloys, and aluminum. Nominal compositions of the principal materials for tubes are given in Table 10:2.

TABLE 10:2
NOMINAL COMPOSITIONS OF SOME ALLOYS USED FOR COLLAPSIBLE TUBES

Composition, %

Pb	Sn	Sb	Cu	Ag	Al
		Lead Alloys			
99.7	—	—	—	—	—
98	[a]	2	—	—	—
97	[a]	3	—	—	—
96.5	—	3	—	0.5	—
96	—	3	—	1	—
		Tin Alloys			
—	99.8	—	—	—	—
—	99.6	—	0.4	—	—
—	99	—	1	—	—
		Aluminum			
—	—	—	—	—	99.99
—	—	—	—	—	99.7

[a] Tin-coated lead tubes are produced from these alloys, the coating being equivalent to 3 to 7% tin.

The production of collapsible tubes is one of the major uses of impact extrusion. The scope of the operation can be judged from a recent release by the Collapsible Tube Manufacturers Council. According to the Council, over one billion units are produced per year. Of these, 52 per cent are used as containers for toothpaste, the remainder being used for packaging shaving creams, medicinal and pharmaceutical products, household and industrial products, and cosmetics.

PRODUCTION OF TUBES

The sequence of operations in fabricating a tube is essentially the same for all materials, although certain modifications are required in using aluminum.[2] These operations are:

- Casting
- Preparing Blank
- Lubricating
- Impact Extrusion
- Trimming and threading the neck
- Enameling and coating
- Capping
- Final inspection and packaging

LEAD AND TIN TUBES

Identical procedures are used in the production of lead and tin tubes and these materials can be discussed together.

Casting and Rolling

Alloys of lead and tin are generally prepared by melting the metals in a kettle and casting the mixture into slabs which are rolled cold to the thickness desired. As the diameter of the blank prepared from the strip is practically the same as that of the finished tube, the metal required to form the walls is obtained by adjusting the thickness of the rolled strip.

As an example, tin-copper tubes are produced by melting Straits tin with 0.5 per cent of copper and casting it into slabs which are $16 \times 4 \times \frac{5}{8}$ inches. The slabs are rolled to a thickness of $\frac{1}{16}$ inch. A gross of tubes $1\frac{1}{16} \times 5\frac{3}{16}$ inches can be prepared from one of these slabs.[3]

As tin is a relatively expensive metal, some tubes are made of tin-coated lead. According to Friden,[4] a tin-coated lead tube is produced from strip prepared by wrapping a thin sheet of pure tin around a carefully cleaned lead slab and rolling this cold to the desired thickness. A firmly adhering coating is obtained which is uniformly distributed during extrusion.

Blanking and Lubrication

Blanks are prepared by punching disks of the desired diameter from the rolled sheets of lead, tin-coated lead, or tin alloy. As mentioned previously, the thickness of the sheet determines the length of the tube. The blanks are generally punched with a tool that forms a slightly dished surface. In an alternate procedure, blanks of these alloys, excluding the tin-coated materials, are poured directly from a kettle onto a casting wheel.

After blanking or casting, the slugs are tumbled for a short time with a suitable lubricant, which may be mineral oil, vegetable oil, a mixture of oil and talc, or various other lubricants. Tumbling time is limited to the minimum required to spread the lubricant uniformly over the surface. Excessive tumbling may cause formation of metal dust when the blanks rub together and this dust may cause pin-holes during extrusion of the tubes.

Extrusion and Trimming

The formation of a collapsible tube requires a combined backward and forward extrusion. The neck is formed by forward extrusion against a spring that limits the distance the metal flows and thus controls the length of the neck. The walls are formed by backward extrusion. The thickness of the blank is calculated to produce a wall somewhat longer than that required for the tube in order to permit trimming the irregular end that results from extrusion.

The rise in temperature resulting from impact extrusion is sufficient to raise the temperature of lead and tin alloys into the hot-working range. Therefore, the tubes formed from these metals are soft and ductile as they come from the press and are ready for trimming and decorating.

The tubes are trimmed to length and the neck is threaded in automatic machines. Tolerances for length must be controlled closely because the tubes are filled by high-speed automatic machinery which is adjusted to close tolerances.

Finishing Operations

The remaining operations are highly mechanized. The trimmed tubes are placed on conveyor lines which carry them to the roller-coating machines. In these machines, a base coat of the desired color is applied. The tubes then pass through an oven and the coatings are dried to the tackiness necessary to retain a good impression. Then they go to the printing machines.

All colors must be printed at the same time on these tubes and a rotary offset process is used with a printing plate for each color. The colors are transferred to a blanket and from the blanket to the tube. After printing, the tubes are baked in an oven to dry the colors and are then ready for capping and final inspection. In some cases, lead tubes require internal protection and are lined with wax. As the inspection methods are essentially the same for all types of tubes they will be discussed after consideration of the production of aluminum tubes.

ALUMINUM TUBES

The use of aluminum tubes is expanding so rapidly that they constitute a large proportion of the current production of collapsible tubes. There is a very good reason for this development. Pierce[5] points out that a gross of lead tubes, $\frac{7}{8}$ by 5 inches, has a shipping weight of about seven pounds whereas a gross of aluminum tubes of the same size has a shipping weight of about one pound. As shipping is a major item of expense in the use of collapsible tubes, producers are changing to aluminum as rapidly as possible.

A pictorial description of the production of aluminum tubes from blank to decorated tube is given by Ott.[6] Briefly the operations are as follows:

Casting and Rolling

Because softness is a requirement of collapsible tubes, it is the general practice to use aluminum that is nominally 99.7% pure. This aluminum is melted and cast into vertical molds. In one installation,[9] the slabs produced are 20 by 15 by $1\frac{1}{2}$ inch. The slabs are hot-rolled in four passes to a thickness of about $\frac{1}{4}$ inch, annealed for two hours at 900 F, and cold-rolled to the desired thickness.

Recently a continuous casting operation has been devised for the preparation of aluminum blanks.[8] In this process, aluminum pigs are melted in an induction furnace and the metal is poured onto a casting wheel on which thin, relatively wide, strip is formed. The strip passes directly to a rolling mill for reduction to the proper thickness.

Blanking and Lubrication

Aluminum slugs are blanked from rolled strip or cut from rod. Some producers of collapsible tubes bypass the casting and rolling operations and purchase blanks from the aluminum producers. The type of blank used depends on the type of nozzle to be produced.[9] Some blanks are flat, others are dished, and either form may have a central hole. The diameter of the blank is related directly to the diameter of the tube; and the length of the tube is controlled by the thickness of the blank. Pierce[5] notes that blanks must be held to close tolerances and, if they are to be pierced, the hole must be centered exactly.

Goldstein[10] has reported a procedure for the production of aluminum blanks by direct casting. He states that these blanks are precisely sized, have controlled granular structure, and are free of pores and other defects.

After being cut from the strip or rod, the blanks must be annealed. One procedure is to anneal at 900 F for two hours.[4] Another is to heat the blanks near the melting point of aluminum for 20 minutes.[5]

After annealing, the blanks are lubricated in a tumbling barrel or by other means. To avoid formation of aluminum powder, which has been blamed for causing pinholes in tubes, the tumbling time is very short, usually ranging from two to six minutes.[11] Various lubricants are used, the most popular being lanolin-base greases or proprietary extreme-pressure lubricants.

Extrusion, Trimming, and Annealing

Although pressures are somewhat higher, aluminum tubes are extruded in exactly the same manner as lead tubes. As the rise in temperature during extrusion is not sufficient to anneal the tubes they are hard where they are removed from the press. After extrusion, the tubes are trimmed to length and threads are either cut or rolled on the necks. The tubes must then be annealed to soften them sufficiently to enable them to function as collapsible tubes.

A number of methods have been used for annealing the tubes. In a muffle furnace with circulating-air, maintained at a temperature of 700 F, it is reported that loads of 200 pounds require two hours after reaching the temperature of the furnace.[12] The same source suggests use of a conveyor furnace having a seven-foot heating chamber and a belt eighteen inches wide With a load of twenty pounds to the foot of length of belt at a temperature of 700 F, satisfactory results can be obtained at economical belt speeds.

A high-speed annealing procedure has been described by Malany.[13] The aluminum tubes are passed on a single-line conveyor through a radiant-burner furnace. As the tubes pass between two banks of these burners, each is heated individually from room temperature to the required annealing temperature of 650 F in less than 30 seconds.

Internal Lacquering, Coating, and Printing

One of the problems that arose in attempting to introduce aluminum collapsible tubes was the reactivity of aluminum with some of the products that can be packaged in lead or tin tubes. In some cases, it was possible to modify the formula to make it less corrosive; in others, inhibitors were used. Thus sodium silicate is used with some shaving creams to prevent attack.[18] It is often difficult or undesirable to modify the product, however, and other means of protection must be used.

Early attempts were made to protect the aluminum with waxes or lacquers, as liners for the tubes. These were not entirely successful, however, because of failure to obtain complete coverage or because the liners cracked during use. Ott[6] points out that the development of ethoxyline (epoxy) resins, which could be processed into highly flexible coatings, greatly expanded the applications of aluminum tubes. These coatings are alkali- and acid-proof, odorless, and tasteless. In addition, their color offers an easy means of inspection for complete coverage. Carr[9] notes that they cannot be used for tubes less than $\frac{5}{8}$ inch in diameter because it is difficult to obtain a spray system small enough to operate below this diameter. In practice, the resin is sprayed into the tube which is then heated to a temperature high enough to cause polycondensation of the resin. This step precedes decorating.

The tubes are then roller-coated with a base of colored enamel and passed through an oven, as described under *Finishing Operations* for lead and tin tubes.

Both internal lacquer and external decoration must remain flexible for the entire life of any tube, regardless of temperature and humidity.

INSPECTION OF COLLAPSIBLE TUBES

According to Elloitt,[14] the major control tests required are:

Composition of metal
Dimensions: length; diameter; thickness of wall, shoulder, membrane, size and shape of nozzle and orifice and threads.
Pinholing
Resistance to bursting
External finish: appearance, weight, continuity, adhesion, and flexibility of coating.
Internal coatings: weight, continuity, adhesion, and flexibility.

Some of these tests are self-explanatory and require no comment. The others are discussed briefly as follows.

Pinholes

According to Zettler,[15] the common method of detecting pinholes is to insert a strong light into the tube in a darkened room. He noted, however, that pinholes can be detected automatically by a combined optical-electrical process. A strong light is projected into the tube as it passes on a conveyor. If there is a pinhole, light passes through it and strikes a photocell. The impulse of the current is amplified sufficiently to operate a relay mechanism that rejects the tube.

Bursting

Air or water pressure is used to determine the resistance of the tube to bursting. As this test is not standardized, requirements are set by agreement between the producer and purchaser.

External Finish

Elliott[14] notes that tests for adhesion vary but all employ folding and unfolding procedures and examination for detachment of the coating.

Thickness of the coating can be determined by weighing samples from the tubes accurately, removing the coating with a solvent, and reweighing the tubes.

Internal Coating

In epoxy-type lacquers, properly heated material has a yellow shade; evenness of color indicates uniformity of coating. Complete coverage is more important, however. Continuity of coating on aluminum tubes can be tested by the use of a saturated solution of mercuric chloride at room temperature. The solution is poured into the tube, allowed to remain about two minutes, and discarded. After being rinsed, the tube is opened for examination. Discontinuities in the lacquer are indicated by a white powdery excrescence.[14]

Continuity of the coating on a tin-coated lead tube can also be determined by a simple spot test. The surface must be washed free of grease. Either concentrated nitric or acetic acid is placed on the tin-coating, the specimen is carefully heated to evaporate the acid, and a solution of 5% potassium iodide is dropped on the treated area. Yellow needles of lead iodide form at pinholes or other discontinuities.[16] The same test can be used to detect contamination of aluminum tubes by lead.

Resistance to Products

Elliott[14] states that there are no standard tests for resistance of tubes to their contents. Usually, filled tubes are exposed at a constant temperature for a specified time and then examined. He suggests, for example, a temperature of 100 F for 50 days. Suitability may be judged by such criteria as no sign of attack, freedom from swelling, freedom from leakage, etc.

APPLICATIONS OF COLLAPSIBLE TUBES

Collapsible tubes are used for packaging a wide variety of products. These include toothpastes, shaving creams, hair creams, hair dyes, shampoos, permanent wave creams, vanishing creams, suntan creams, paints, lubricating oils, greases, duplicating inks, adhesives, furniture polishes, shoe polishes, pharmaceuticals, cheese products, and various foodstuffs.[6,9,17]

REFERENCES

1. *Glossary of Packaging Terms*, British Standard No. 3130, 1959.
2. Collapsible Tube Production; Impact Extrusion, *Mass Production*, 32, June 1956:84.
3. Producing Collapsible Tubes, *Machinery* (London), 44:573, (1934).
4. Friden, J. H., Impact Extrusion of Lead, *Metals Handbook*, 1948 edn., 949; American Society for Metals.
5. Pierce, H., Impact Extrusion Methods at Western Aluminum Plant, *Light Metal Age*, August 1948:18.
6. Ott, G. H., The Aluminum Collapsible Tube, Its Development and Use, *Aluminium Suisse*, November 1952:185.

7. German Collapsible Tube and Extrusion Industries, *B.I.O.S. Final Report No. 438*, 1946.
8. Continuous Casting, Impact Extrusion Spark Revolution in Aluminum Partmaking, *Steel*, January 12, 1959, 70.
9. CARR, R. W. D., Aluminum Collapsible Tubes for Packaging, Paper in Symposium on *Aluminum in Packaging*, The Aluminum Development Association, June 12, 1958.
10. GOLDSTEIN, M. M., New Development in Impact Extrusion Field Extends Process Control, *Industrial Heating*, January 1957:75.
11. Impact Extrusion in Germany, *Machinery* (London), October 23, 1947:460.
12. Impact Extrusion in Practice, *Light Metals*, 5:177 (1942).
13. MALANY, R., Mechanized Annealing of Aluminum Collapsible Tubes, *Modern Metals*, January 1955:73.
14. ELLIOT, E., Aluminum Collapsible Tubes, Corrosion and Production Considerations, *Corrosion Technology*, 7, June 1960:171.
15. ZETTLER, E., The Present Position of the Production and Use of Aluminum Collapsible Tubes, *Metallwirtschaft*, 18:881 (1939).
16. FEIGL, F., *Spot Tests*, Elszevier Publishing Company, 1954.
17. D'ANTONIO, A. A., Packaging Cosmetics in Collapsible Metals Tubes Increases, *American Perfumer*, 76, August 1961:52.
18. Sodium Silicate Inhibits Corrosion of Collapsible Tubes by Soap, *Scientific American*, 143:478 (1930).

Chapter 11

OTHER NONFERROUS METALS

The most significant advances made in the application of impact extrusion have been in the production of parts from aluminum and aluminum alloys. The process is gradually being expanded to other nonferrous metals, however, and some of the developments are discussed in the following sections.

COPPER AND COPPER ALLOYS

Although copper was forward-extruded into tubes at least thirty-five years ago, application of impact extrusion to the copper alloys has developed slowly. Increasing interest in the process within the past few years has resulted in some work on these materials and experimental work has shown that a number of the standard copper alloys can be backward-extruded. Applications are developing in a number of fields.

Fitness[1] reports that copper, high-conductivity copper alloys, such as copper-silver alloys, and brasses containing up to 20 per cent of zinc can be backward-extruded readily. He notes that the 30 and 35 per cent zinc alloys require higher extrusion pressures but can also be extruded. Howard[2] agrees with these conclusions and adds that the 40%-zinc brass cannot be extruded satisfactorily. Others have reported the backward extrusion of beryllium copper; a copper-nickel alloy containing 10 per cent of nickel; and a nickel silver containing 18 per cent of nickel.

The compositions of some of the alloys that have been impact-extruded either on a commercial or on an experimental basis are given in Table 11 : 1. In general, the starting blanks are annealed before extrusion. Annealed properties of the alloys

TABLE 11:1

NOMINAL COMPOSITIONS OF SOME COPPER ALLOYS USED FOR IMPACT EXTRUSION

Designation	*Common Name*	*Composition, %*		
		Cu	Zn	Other
Copper No. 102	Oxygen-Free Copper	99.96	—	—
Copper No. 110	Electrolytic Tough-Pitch Copper	99.92	—	0.04 oxygen
Copper Alloy No. 172	Beryllium Copper	Rem		2.0 Be, 0.2 Co
Copper Alloy No. 210	Gilding, 95%	95	5	—
Copper Alloy No. 220	Commercial Bronze, 90%	90	10	—
Copper Alloy No. 230	Red Brass, 85%	85	15	—
Copper Alloy No. 240	Low Brass, 80%	80	20	—
Copper Alloy No. 260	Cartridge Brass, 70%	70	30	—
Copper Alloy No. 270	Yellow Brass, 65%	65	35	—
Copper Alloy No. 752	Nickel Silver, 65-18	65	17	18 Ni
Copper Alloy No. 706	Cupro-Nickel, 10%	Rem	—	10 Ni, 1.25 Fe

listed in Table 11:1 are given in Table 11:2 in the order of increasing hardness. As the strength and the hardness are indications of relative extrudability in a single alloy system, this table indicates that, of the materials listed, oxygen-free copper is the easiest to extrude, and beryllium copper, the most difficult.

These deductions are confirmed to some degree by the data

TABLE 11:2

PROPERTIES OF ANNEALED COPPER ALLOYS ARRANGED IN ORDER OF INCREASING HARDNESS*

Alloy	*Yield Strength (0.5% extn.) psi*	*Tensile Strength psi*	*Elongation (2 in.)*	*Shear Strength psi*	*Rockwell F Hardness*
Oxygen-Free Copper	6,500	32,000	40	22,000	30
Electrolytic Tough-Pitch Copper	10,000	32,000	45	22,000	40
Gilding, 95%	12,000	36,000	43	30,000	53
Commercial Bronze, 90%	12,500	39,000	43	30,000	59
Red Brass, 85%	14,000	42,000	46	32,000	63
Cupro-Nickel, 10%	16,000	44,000	42	—	64
Low Brass, 80%	16,000	46,000	49	32,000	65
Cartridge Brass, 70%	16,500	48,000	60	34,000	66
Yellow Brass, 65%	18,000	49,000	59	34,000	68
Nickel Silver, 65-18	25,000	58,000	40	—	85
Beryllium Copper	—	72,000	20	—	94

* Average values—annealed properties vary with size of grain

TABLE 11:3

APPROXIMATE EXTRUSION PRESSURES FOR SOME ANNEALED COPPER ALLOYS

Alloy	*Extrusion Pressure* (1000 *psi*) *for Reduction of Area of*						*Reference*
	50%	70%	80%	85%	87%	90%	
Copper (99.9%)	80	112	150	—	—	—	
Copper (99.6%)	—	157	—	—	—	—	
Copper (99.5%)	—	144	168	—	—	—	5
Copper	—	—	197	—	258	—	2
Copper	—	—	—	233	—	270	1
Cartridge Brass, 70%	130	190	246	—	—	—	3
Yellow Brass, 65%	—	219	269	—	—	—	5
Nickel Silver, 65-18	130	193	251	—	—	—	3
Beryllium Copper	168	246	—	—	—	—	3

included in Table 11:3, which lists the extrusion pressures determined by a number of investigators during the backward extrusion of round-bottomed shells. Although the table is incomplete, there is a definite increase in the extrusion pressure with

TABLE 11:4

TYPICAL PROPERTIES OF SOME WROUGHT COPPER ALLOYS

Alloy	*Condition*	*Yield Strength* (0.5% *extn.*) *psi*	*Tensile Strength psi*	*Rockwell Hardness*
Electrolytic Tough-Pitch Copper	Annealed*	10,000	32,000	F40
	Half Hard	36,000	42,000	F84
	Hard	45,000	52,000	F90
Commercial Bronze, 10%	Annealed*	12,500	39,000	F59
	Half Hard	45,000	52,000	B58
	Hard	54,000	61,000	B70
Low Brass, 20%	Annealed*	16,000	46,000	F65
	Half Hard	50,000	61,000	B70
	Hard	59,000	74,000	B82
Cartridge Brass, 30%	Annealed*	16,500	48,500	F66
	Half Hard	52,000	62,000	B70
	Hard	63,000	76,000	B82
Nickel Silver, 65-18	Annealed*	25,000	58,000	B40
	Half Hard	62,000	74,000	B83
	Hard	74,000	85,000	B87

* Average values; annealed properties vary with size of grain

increasing hardness of the material. Roughly twice as much pressure is required to extrude beryllium copper as to extrude standard copper at equivalent reductions of area.

The impact extrusion of copper is a cold-working operation; therefore, the hardness and strength of a part are increased during the operation. Unfortunately, few data have been published on the improvement in properties that result from impact extrusion. Table 11:4, however, shows the strengths developed in other cold-working operations and thus gives an indication of the improvement that can be expected.

Fukui and his associates[5] determined the change in hardness resulting from the backward extrusion of copper and yellow brass into round shells. Their results, given in Table 11:5, show that considerable increase in hardness occurs if the materials are in the annealed condition before extrusion. Only a slight increase occurs if the starting material is in the worked condition.

TABLE 11:5

IMPROVEMENT IN HARDNESS OF COPPER ALLOYS REDUCED 80% BY BACKWARD EXTRUSION*

Alloy	*Condition*	*Initial Vickers Hardness*	*Final Vickers Hardness*		
			Upper Wall	Lower Wall	Bottom
Copper	Annealed 1300 F, 1 hr, furnace cooled	38	120	124	127
Copper	Hot extruded, cold upset	110	128	129	133
Yellow Brass	Annealed 1025 F, 1 hr, furnace cooled	55	195	223	223

* Fukui, Kudo, and Seino[5]

COMMERCIAL PRODUCTS

There seem to be no published specifications for parts produced from copper-base materials. Fitness[1] notes that the ratio of length to diameter for backward extrusion is limited to about 5 to 1; and the maximum reduction of area, to about 95 per cent for copper, 75 per cent for brasses containing up to 20 per cent of zinc, and 60 per cent for cartridge brass.

Tolerances that can be held on small parts are:

Wall-thickness ±0.003
Diameter ±0.003
Finish can be held to 63 microinch rms[6]

In forward extrusion there is much greater latitude in the ratio of length to diameter. Copper tubes that have either straight or tapered walls are extruded with open or closed bottoms. Tubes 14 inches long $\frac{3}{16}$ to $\frac{3}{8}$ inch O.D. with walls ranging from 0.004 to 0.010 inch thick, are produced from preformed blanks with a reduction of area of 94 per cent.

FIG. 11:1. TRANSISTOR HEAT SINK PRODUCED FROM OXYGEN-FREE COPPER (*courtesy* Steel Industries, Inc.)

APPLICATIONS

The excellent electrical properties of copper have been a major reason for the use of this material. It is not surprising therefore, that advantage has been taken of impact extrusion to produce parts that can be used in electrical applications.

Because it has the best ductility of the coppers, oxygen-free copper is used more widely than any other copper-base material. Parts produced from this copper for the electronic industry include rectifier bases, terminals, and transistor heat sinks of various forms, one of which is illustrated in Figure 11:1. Thermostat cups are also being backward-extruded from this material. Oxygen-free copper has been forward-extruded into tubes for use in various heat-exchangers. A use not related to its electrical and thermal conductivity is the backward extrusion of a thin-walled rectangular can for a pocket lighter.

Electrolytic tough-pitch copper, tellurium copper, and low-phosphorus copper are being extruded into various small shells.

Crampton[7] discussed the production of small cartridge cases from 70/30 brass by forward extrusion. During the course of a survey on impact extrusion, however, it was reported that the process was not sufficiently economical to justify use of it for this application.

Some of the other copper alloys are being used for the production of engineering components where strength rather than electrical or thermal properties is the criterion for service. Among these products are:

1. A *valve body*, which is being impact-extruded from brass in a single operation. This part was formerly produced by machining. Conversion to impact extrusion reduced the material required by 50 per cent and considerably increased the rate of production.

2. A *pay-telephone adapter*, which is being extruded from nickel silver, 65-18. It is claimed that the part, after extrusion, has mechanical properties approaching those of steel.[8]

3. A *piston* for use in a valve operating at high pressure, which is being extruded from an aluminum bronze containing 5 per cent nickel.[8] The part is heat-treated to obtain the high strength required for satisfactory operation of the valve.

Forward extrusion is being used to produce plain and ribbed tubes of various alloys including copper-nickel, 10%.

MAGNESIUM AND MAGNESIUM ALLOYS

The impact extrusion of magnesium differs in one respect from that of the metals discussed previously. As the structure of magnesium is such that it cannot be worked readily in the cold, the blanks must be warmed. Generally, temperatures in the range of 450 to 700 F are used for impact extrusion.

Originally the process was confined to the production of simple shapes from unalloyed magnesium. It has been extended, however, to some of the alloys, of which the nominal compositions are given in Table 11:6.

TABLE 11:6

NOMINAL COMPOSITIONS OF SOME MAGNESIUM ALLOYS USED FOR IMPACT EXTRUSION

Alloy	*Composition, %*			
	Al	Zn	Mn	Zr
AZ31B	3.0	1.0	—	—
AZ61A	6.5	1.0	—	—
AZ80A	8.5	0.5	—	—
M1A	—	—	1.2	—
ZK60A	—	5.5	—	0.5

Although the blanks are heated and are therefore softened to some extent, the mechanical properties at room temperature will give at least an indication of their relative strengths and hardnesses. In Table 11 : 7, the alloys are listed in order of increasing hardness and, as shown in Table 11 : 8, the required extrusion pressures increase with the hardness. It is noteworthy that because of the use of heated slugs, the range of extrusion pressures required for these alloys is considerably lower than would be

TABLE 11:7

PROPERTIES OF ANNEALED MAGNESIUM ALLOYS ARRANGED IN ORDER OF INCREASING HARDNESS

Alloy	*Yield Strength (0.2% offset), psi*	*Tensile Strength psi*	*Elongation (2 in.) %*	*Compressive Yield Strength psi*	*Shear Strength psi*	*Brinell Hardness*
M1A	20,000	34,000	9	12,000	18,000	44
AZ31B	26,000	37,000	12	15,000	19,000	49
AZ61A	30,000	44,000	14	19,000	19,000	60
AZ80A	32,000	48,000	12	—	22,000	62
ZK60A	38,000	49,000	12	33,000	24,000	75

TABLE 11:8

PRESSURES FOR REQUIRED IMPACT EXTRUSION OF VARIOUS MAGNESIUM ALLOYS*

(85% Reduction of Area)

Alloy	*Extrusion Pressure (tons per square inch) at*						
	450F	500F	550F	600F	650F	700F	750F
AZ31B	33	33	30	27	26	25	23
AZ61A	35	34	33	32	31	30	29
AZ80A	36	35	34	33	32	31	30
ZK60A	34	33	32	31	29	27	26

* Dow Metal Products Company[9]

expected on the basis of the degree of hardness at room temperature.

It will be recalled that the impact extrusion of aluminum alloys resulted in a considerable improvement in mechanical properties, because the operation was cold-working. With magnesium, impact extrusion is probably a borderline operation between cold- and hot-working and, as a result there may or may not be an improvement in properties. As shown in Table 11:9, limited data indicate that there was a moderate increase in strength in one alloy and a moderate decrease in the other.

TABLE 11:9
PROPERTIES OBTAINED ON IMPACT-EXTRUDED MAGNESIUM PARTS*

Alloy	*Yield Strength psi*		*Tensile Strength psi*		*Compressive Yield Strength psi*	
	Annealed	Impact Extruded	Annealed	Impact Extruded	Annealed	Impact Extruded
AZ31B	22,000	28,000	37,000	43,000	16,000	23,000
ZK60A	38,000	34,000	49,000	46,000	33,000	28,000

* Everhart[10]

COMMERCIAL PRODUCTS

The commercial limit of the ratio of length to diameter in backward extrusion is about 8 to 1, whereas in forward extrusion, the limiting factors are the clearance required and the size of blank that is practical. Parts with walls ranging from 0.006 to 0.25 inch thick and diameters of 1 to 6½ inches are produced commercially.[11] Tolerances depend on the shape of the part, ratio of length to diameter, and characteristics of the external press. Typical tolerances are given in Table 11:10.

TABLE 11:10
TYPICAL TOLERANCES FOR IMPACT-EXTRUDED MAGNESIUM PARTS*

Dimension[a]	*Tolerance, inch*
Diameter	±0.002 per inch of diameter
Bottom Thickness	±0.005 on all sizes
Wall Thickness	
0.020-0.029	±0.002
0.030-0.044	±0.003
0.045-0.059	±0.004
0.060-0.100	±0.005

[a] For ratios of length to diameter, 6 to 1
* Dow Metal Products Company[9]

APPLICATIONS

Generally, the shapes produced from magnesium alloys are more restricted than those produced from aluminum. Most parts are can shaped, with or without ribs, bosses, etc. Cross sections are usually circular, square, or rectangular. Parts are used in pyrotechnics, batteries, and some electronic applications.[12] Examples of parts in magnesium alloy are shown in Figure 11:2.

An interesting product is a triangular bracket, produced by impact extrusion of a rectangular can which is then cut into two brackets. This part was originally produced by blanking from sheet stock, welding the corners, relieving stress, and sanding.

Another part converted to impact extrusion is an electronic shield which is produced as a square-base can from the AZ31B alloy. This part was formerly produced by deep-drawing in five steps. It is now impact-extruded in one step and requires only trimming to length.

NICKEL AND NICKEL ALLOYS

Little work on the impact extrusion of nickel and its alloys has been reported. Tilsley and Howard[13] state that pure annealed nickel can be extruded successfully with reductions of area up to 80 per cent. According to these investigators, the extrusion pressures required range from 290,000 to 335,000 *psi* and the best surface conditions and lowest pressures result if the blanks are copper-plated and lubricated with chlorinated mineral oils.

TABLE 11:11
PROPERTIES OF NICKEL AND A NICKEL ALLOY USED FOR IMPACT EXTRUSION

Metal	*Condition*	*Yield Strength (0.2% offset) psi*	*Tensile Strength psi*	*Elongation, (2 in.), %*	*Rockwell B Hardness*
Nickel (99.5%)	Annealed	15,000-30,000	60,000-80,000	50-35	40-65
	Cold Worked	70,000-115,000	90,000-130,000	15-2	96 min
70% Ni/30% Cu	Annealed	25,000-30,000	70,000-80,000	50-35	60-75
	Cold Worked	90,000-130,000	100,000-140,000	15-2	98 min

FIG. 11:2. MISCELLANEOUS PARTS IMPACT-EXTRUDED FROM MAGNESIUM ALLOYS (*courtesy* Dow Metal Products Company)

These extrusion pressures are in the range required for steel extrusions. It is possible that development of a suitable coating, such as the phosphate coating employed for steel, will improve the extrudability and reduce the pressure requirements. It has been reported that the alloy of 70 per cent nickel with 30 per cent copper has also been impact-extruded.

The ratio of length to diameter for nickel parts is limited to about 4 to 1 for backward extrusion. Longer parts can be produced by forward extrusion. Properties of nickel and the 70/30 nickel-copper alloy in the annealed condition and after cold-working are given in Table 11:11.

The 70/30 nickel-copper alloy has been impact-extruded into turbine blades.[14] In this development, four turbine blades are backward-extruded simultaneously from heated blanks (temperature not stated). An example was given of an integral root-type blade which when extruded was 0.6 inch wide and 0.205 inch (maximum) thick.

TITANIUM

Commercially pure titanium can be backward- and forward-extruded without difficulty provided the surface is suitably treated. High extrusion pressures are required. The process has not been applied to titanium alloys.

Commercial titanium is usually graded by tensile or yield strength, although specifications call for definite limits on the content of impurities in the various grades. As annealed, the purest material has the lowest yield strength. Table 11:12 lists the three grades of commercial titanium used most frequently, together with typical annealed properties.

TABLE 11:12

NOMINAL COMPOSITIONS AND ANNEALED PROPERTIES OF COMMERCIAL TITANIUM USED FOR IMPACT EXTRUSION

Alloy	*AMS Designation*	*Typical Properties*		
		Yield Strength (0.2% offset), psi	Tensile Strength psi	Elongation (2 in.) %
99.2% Ti	4902	40,000	59,000	28
99.0% Ti	4900	63,000	79,000	27
99.0% Ti	4901	80,000	95,000	25

A surface treatment developed by Battelle Memorial Institute has been used successfully in the extrusion of titanium. This treatment consists of a fluoride-phosphate coating with a lubricant of graphite and gum resin.[15] According to Sabroff and his associates,[16] grade AMS 4900 titanium can be backward-extruded with reductions of area up to 50 per cent if this lubricant is used. They reported an extrusion pressure of 394,000 *psi* for a reduction of 50 per cent.

This agrees with the value reported for a similar reduction of area by Ridings and Angus.[3] Although they used a lubricant consisting of colloidal graphite and oil, their results indicated an extrusion pressure of 392,000 *psi*. These researchers also investigated the use of heated blanks. They found that the extrusion pressure could be reduced by 55 per cent when the blanks were preheated to 1100 F.

Using the Battelle lubricant, Sabroff and his associates[16] were able to forward-extrude parts with a reduction of area of 60 per cent. At this reduction, extrusion pressures were 169,000 *psi* for a solid extruded part and 148,000 *psi* for a hollow part.

Backward-extruded shells were produced in sizes up to 3 inches long with walls as low as 0.22 inch thick; in forward-extruded tubes this thickness was as low as 0.15 inch. In these tests, the surface finish ranged from 30 to 60 microinches *rms* which is equivalent to a smooth turned finish.

The Battelle group also developed a method of backward-extruding hexagonal nuts, using AMS 4902 titanium. It was reported that the backward extrusion of these nuts reduced the cost of production by almost 50 per cent as compared with conventional methods.[17]

TABLE 11:13
NOMINAL COMPOSITIONS OF SOME ZINC ALLOYS USED FOR IMPACT EXTRUSION

Alloy	*Composition, %*						
	Zn	Pb	Cd	Cu	Fe	Mg	Al
Pure Zinc	99.99	—	—	—	—	—	—
Commercial Zinc	Rem	0.8	0.06	0.005 max	0.012 max	—	—
Zinc-Copper Alloy	Rem	—	—	1.0	—	—	—
Zinc-Cadmium Alloy	Rem	—	0.6	—	—	—	—
Zinc-Aluminum-Magnesium-Alloy	Rem	—	—	—	—	0.04	4.0

ZINC AND ZINC ALLOYS

Pure and commercial zinc have been used for many years in the production of impact-extruded parts. A limited number of alloys have also been used. Nominal compositions of these materials are given in Table 11:13.

According to Askew,[18] backward extrusion should be performed on blanks heated to about 350-450 F in order to reduce the working pressure. As shown in Table 11:14, the extrusion pressures can be relatively high if the operation is performed at room temperature on cold blanks.[19] Others, however, have extruded high-purity zinc at room temperature, depending on the heat of extrusion to warm the tools and reduce the required extrusion pressure.

TABLE 11:14

APPROXIMATE EXTRUSION PRESSURES FOR ZINC

Condition	*Extrusion Pressure (psi) for Reduction of Area of*			*Reference*
	60%	85%	95%	
As Rolled	122,000	—	203,000	19
Annealed 300 F, 70 min	124,000	—	221,000	19
—	157,000	269,000	—	2

No improvement in strength results from the impact extrusion of zinc; the parts are soft as they come from the press. For example, Mabb[20] reports that the following measurements for hardness were obtained on a backward-extruded shell:

Part Tested	*Diamond Pyramid Hardness*
Blank	38
Extruded Shell	
Top of wall	37
Middle of wall	34
Bottom of wall	34
Bottom of shell	34

In general, commercial parts are produced in sizes ranging from 0.5 to 2.5 inches in diameter. Lengths of backward extruded parts range from 2.5 to 3 inches for parts of small diameter and up to 8 inches for those of larger diameters. Wall-

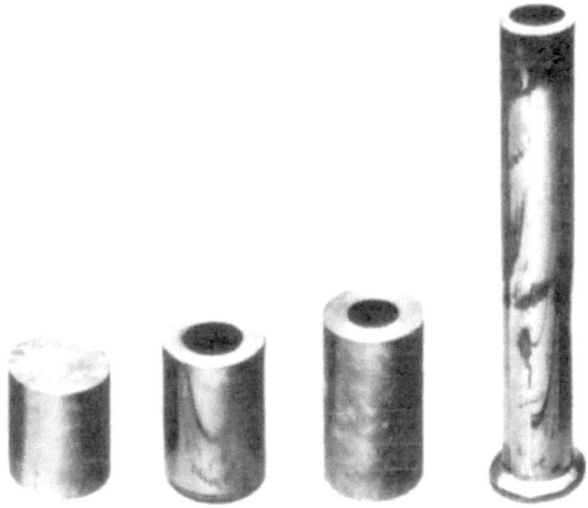

FIG. 11:4. BERYLLIUM TUBE PRODUCED BY IMPACT EXTRUSION
Left to right: Compact; backward-extruded cup; machined cylinder; forward-extruded tube

FIG. 11:3. FLANGED PRESSURE TUBES OF ZIRCALOY-2, PRODUCED BY FORWARD EXTRUSION, FOR PRESSURIZED WATER NUCLEAR REACTOR (*courtesy* Bridgeport Brass Company, Division of National Distillers and Chemical Corporation)

thickness varies with the diameter. The minimum thickness ranges from 0.014 inch for parts of small diameter to 0.020 inch for larger parts.

The major use of zinc for impact extrusions is in the production of cans for dry cells. These may be produced from commercial zinc or the zinc-cadmium alloy.

Various shells, of round, square, rectangular, and other shapes, are produced from the alloys listed in Table 11:13. Generally, the parts are limited to rather simple shapes. These cans are preferred over aluminum cans for parts in automobile radios because zinc has low resonance properties and it can be soldered readily. Zinc shells are used also as shields in electronic equipment and for noise filters.

ZIRCONIUM AND ZIRCONIUM ALLOYS

Commercial zirconium and Zircaloy 2 have been successfully impact-extruded. The extrusion of other zirconium base materials has not been reported.

According to Quadt,[21] zirconium of commercial grade (containing hafnium) and reactor grade (hafnium-free) can be forward-extruded readily. Tubes were produced from annealed blanks at 550 F with a reduction of area of 75 per cent. The tensile strength of the extruded part was roughly twice that of the annealed blank.

Ridings and Angus[3] investigated the backward extrusion of Zircaloy 2 (reactor grade zirconium containing 1.5 per cent of tin). Using a cold blank, formation of a cup, with a reduction of area of 40 per cent, required an extrusion pressure of 388,000 *psi*. By heating the blank to 600 F, they were able to produce a cup having a reduction of area of 82 per cent with the same extrusion pressure as that required for a 40 per cent reduction at room temperature.

Goodwin and Tombaugh[22] also backward-extruded Zircaloy 2 blanks into shells. They used copper-jacketed blanks heated to the range 800 to 1400 F for their experiments and reported successful extrusion of cups at all temperatures above 800 F.

Quadt[23] forward-extruded Zircaloy 2 into tubes. Annealed blanks were used but the conditions were not stated.

Parts produced from zirconium and Zircaloy 2 are used in the field of nuclear energy, but as many applications are classified, there has been little information disclosed about them. They are used as components in moderators, for cladding nuclear fuels, and for boilers and other pressure vessels. According to Quadt, a major advantage of the cold extrusion of zirconium is the recovery of 85 per cent of the weight of the original billet in the final shape. In the use of a metal as expensive as zirconium, this ability to recover most of the metal in the finished part means that costs can be held to the minimum.

FIG. 11:5. MOLYBDENUM TUBE PRODUCED BY BACKWARD-EXTRUDING SINTERED COMPACT INTO A CUP, AND THEN FORWARD-EXTRUDING TUBE AT 1000 F

MISCELLANEOUS METALS

A number of other metals have been impact-extruded but little information is available about them. They do serve, however, to indicate the possibilities of the process and to enable one to predict that impact extrusion will be applied to many more of the ductile metals in the future.

Beryllium

Beryllium has been forward-extruded into tubes at temperatures below 800 F. According to Quadt,[21] there were no unusual problems in producing the tubes.

Columbium (Niobium)

Cast columbium of high purity (Brinell hardness 100) can be extruded readily.[21] Cups were backward-extruded at room temperature into preforms for the forward extrusion of tubes. The tubes were produced at 800 F.

Molybdenum

Arc-cast molybdenum can be impact-extruded if the granular structure is refined.[24] The process involved working the casting by hot extrusion and hot rolling, forming a cup by backward extrusion at 800 F, and forward-extruding the cup into a tube at 800 F. The reduction of area was about 75 per cent. The finished tube was 16 inches long and 1.5 inches in outside diameter, with a wall 0.125 inch thick.

Quadt[21] states that both arc-cast and sintered molybdenum can be used. The practical minimum wall-thickness is about 0.030 inch. He noted that the cold extrusion of molybdenum alloys has not been developed. He mentioned that the forward extrusion of molybdenum is particularly advantageous because of the difficulty of forming tubes by conventional methods of hot extrusion.

Silver

Fine silver has been backward-extruded into electronic parts.[12] No details are available.

Tantalum

High-purity tantalum (Brinell hardness 100) can be forward-extruded into tubes.[21] However, lower grade material (Brinell hardness 160) cannot be successfully cold-extruded because of the excessive extrusion pressures required.

Uranium

Uranium can be forward-extruded at approximately 800 F into thick-walled tubes.[21] The major problems are reported to be related to oxidation and to the development of suitable lubricants.

REFERENCES

1. Fitness, A. D., Copper and Impact Extrusion, *Copper*, No. 12, Summer 1961:20.
2. Howard, F., Designing with Impact Extruded Components, *Engineering Materials and Design* 1, October 1958:26.

3. RIDINGS, J. R., and ANGUS, N. S., Experiments on the Reverse Impact Extrusion of Certain Metals, *Jour. Inst. Metals*, 90:107 (1961).
4. LANDBERG, P., Some Remarks on Impact Extrusion, Microtechnic, 12:301 (1958).
5. FUKUI, S., KUDO, H., and SEINO, J., Study on Impact Extrusion Method for Nonferrous Metals, *Report Inst. Science and Technology, University Tokyo*, 11:91 (1957).
6. Steel Industries, Inc., private communication, July 1962.
7. CRAMPTON, D. K., Extrusion of Metals, *Trans. Amer. Soc. Metals*, 25:55 (1937).
8. Metal Choice, *Precision Metal Molding*, 19, October, 1961:43.
9. Dow Metal Products Co., private communication, July 1962.
10. EVERHART, J. L., Impact (Cold) Extruded Parts, *Materials and Methods*, August 1955:111.
11. PATTON, T. L., Designing Magnesium Impacts, Machine Design, 26 February 1954:124.
12. Cutting Costs with Impact Extrusion, *Steel*, 140, May 6, 1957: 85.
13. TILSLEY, R., and HOWARD, P., Cold Extrusion of Ferrous and Non-ferrous Metals, *Machinery* (London), 93, December 7, 1960: 1286.
14. Extrusion of Nickel Alloys, *Machinery* (London), February 6, 1947:163.
15. SABROFF, A. M., SANNICANDRO, R. A., and FROST, P. D., A Comparison of Lubricants and Coatings for Cold Extruding Titanium, *ASLE Trans.*, 3, April 1960:61.
16. SABROFF, A. M., SANNICANDRO, R. A., and FROST, P. D., Cold Extrusion of Titanium, *Tool Engineer*, 41, July 1958:84.
17. Cold Extrusion of Titanium, *Light Metal Age*, 19, 1961, 12.
18. ASKEW, W. G., Progress in Impact Extrusion, *Metal Industry*, 54:65 (1939).
19. GOKYÛ, I., SUZUKI, H., and SHINJI, K., Studies on Impact Extrusion (7th Report) On Cold Extrusion Force, *Nippon Kinzuko Gakki-Si*, 19, July 1955:18.
20. MABB, P., Impact Extrusion, *Machinery* (London), 76, February 9, 1950:192.
21. QUADT, R., Cold Extrusion for High Temperature Metals, *Western Metalworking*, November 1958:45.
22. GOODWIN, J. G., and TOMBAUGH, R. W., Back Extrusion of Heavy Zircaloy 2 Cups, *Mechanical Engineering*, 79:936 (1957).
23. QUADT, R. A., Precision Cold Extrusion of Metals, *Metal Progress*, May, 1960:121.
24. Forming Molybdenum Tubing, *Steel*, August 11, 1958:821.

Chapter 12

FERROUS METALS

The fastest growing area in the development of impact extrusion is the production of steel parts. In the ferrous industry, the process is known almost universally as *cold extrusion.*

Cold extrusion of steel was started in Germany about 25 years ago. During World War II, the process was used to produce ordnance components. After the war, it was revealed to the Allies and development was started in the United States. The early work was performed by the Heintz Manufacturing Company under contract with the Government to develop ordnance items, mainly artillery projectiles.

Applications of cold extrusion developed slowly during the late 1940's and early 1950's. It is estimated that in 1952, 90 per cent of cold-extruded steel parts were produced for military purposes. By 1962, however, the process had been accepted widely with 90 per cent of the parts being produced for commercial applications and only ten per cent going into military items.

The great advantage of cold extrusion in reducing both costs and losses in scrap is finally being recognized by the mass-production industries, notably the automotive industry. As an example, one producer of piston pins is saving about 2000 tons of steel a year because of conversion from a screw-machine product to a cold-extruded part. Feldmann[1] estimates that approximately 150 different parts are being produced commercially by cold extrusion and many others are in experimental production.

MATERIALS

It appears desirable to begin the discussion of the cold extrusion of steel with some comments on the materials required. In cold extrusion, the metal is subjected to heavy plastic working and the material must be sufficiently ductile to withstand this working without rupture. In addition, the metal resists the deforming forces and the resistance increases with the degree of working. It is desirable that the increase in strength resulting from working be gradual, otherwise the pressure on the surfaces of the tools may build up rapidly and exceed the load-bearing capacities of the tool steels.

For a given structure, both work-hardening characteristics and ductility are closely related to the composition of the steel. Increasing the content of carbon and alloying elements increases resistance to deformation. Assuming that the steel has been heat-treated to obtain the best structure for plastic deformation, maximum deformation can be achieved with minimum content of alloy. Therefore the most suitable steel is one that contains less than 0.10 per cent of carbon. Such a steel is capable of considerable reductions of area and can be formed into complex shapes. According to Okell,[2] increasing the carbon to 0.2 per cent causes increasing difficulty in extrusion and restricts the reduction of area obtainable in a single stage. Commercial extrusion ceases to be practical when the carbon content reaches 0.35 per cent because of excessive wear on the tools.

Okell notes also that a suitable structure can be assumed to be either ferrite-pearlite or free carbide in ferrite, depending on the annealing treatment. The ferrite constituent in a plain carbon steel is soft and most of the deformation occurs in this constituent. Carbide in either form tends to restrict flow but the ferrite matrix is sufficiently ductile to overcome the restriction.

The effects of individual elements, therefore, depend on the manner in which they occur in the steel. As the carbide is brittle and takes little part in the deformation, additional elements that form alloy carbides will have little influence on extrusion. On the other hand, an element that remains in solid solution in the ferrite strengthens the ferrite and has a directly unfavorable effect on extrusion. Thus strong carbide formers, such as chromium, are not so harmful to the process as are ferrite formers, such as nickel. Elements, such as manganese, which divide between the ferrite and the carbide, are intermediate in their effects. Okell states that, at the present stage of development, commercial limits for steels to be used for high production rates are about 0.75 per cent nickel, 1.5 per cent chromium, 1.5 per cent manganese, and 0.5 per cent molybdenum.

In general, Verson[3] agrees with the comments of Okell but he is a bit more specific in his recommendations. He remarks that the content of manganese and silicon also influences the extrudability of the carbon steels. These steels should contain a maximum of 0.30 per cent of silicon and, in fully aluminum-killed steels, the content of silicon should be held in the range of 0.15 to 0.20 per cent. In steels containing less than 0.12 per cent or carbon, the content of manganese can be as high as 1.35 per cent. In higher-carbon steels, manganese should be held in the range of 0.60 to 0.80 per cent; higher manganese increases the rate of hardening and thus raises the extrusion pressure.

In discussing the role of alloying elements, Verson notes that the extrusion pressure increases with the content of nickel. He states that chromium in the 5100 series steels, and chromium-nickel-molybdenum in the 8600 series, cause only slight increases in the extrusion pressure.

A review of numerous articles reporting American practice, together with the results of a questionnaire, sent to various producers, indicate that the steels listed in Table 12:1 are being used for the commercial production of various parts. In general, the compositions agree with the suggestions of Okell and Verson; an exception is the 12%-chromium steel. Undoubtedly, other compositions are also being used. Selection of the specific grade depends on the properties required in the finished part.

Many other steels have been either backward- or forward-extruded experimentally, including carbon steels containing up to 1.20 per cent of carbon, and various chromium and chromium-nickel stainless steels.

This experimental work indicates that there is nothing to prevent extrusion of highly alloyed materials although it may be necessary to modify the process to keep the extrusion pressures within limits that can be handled by the available tool materials. For example, it is quite common in the production of ferrous parts to break the operation into several steps, annealing and relubricating between them. This permits the use of smaller reductions of area in each step than is customary in the impact extrusion of nonferrous metals. In some plants it is the practice to limit the first stage to a reduction of area of not more than

TABLE 12:1
STEELS THAT HAVE BEEN USED COMMERCIALLY FOR COLD EXTRUSION

SAE Number	*Nominal Composition, %*						
	C	Mn	Si	Ni	Cr	Mo	V
1008	0.010 max	0.25-0.50	0.10 max	—	—	—	—
1010	0.08-0.13	0.30-0.60	0.10 max	—	—	—	—
1012	0.10-0.15	0.30-0.60	0.10 max	—	—	—	—
1015	0.13-0.18	0.30-0.60	0.10-0.20	—	—	—	—
1016	0.13-0.18	0.60-0.90	0.10-0.20	—	—	—	—
1017	0.15-0.20	0.30-0.60	0.10-0.20	—	—	—	—
1018	0.15-0.20	0.60-0.90	0.10-0.20	—	—	—	—
1020	0.18-0.23	0.30-0.60	0.10-0.20	—	—	—	—
1021	0.18-0.23	0.60-0.90	0.10-0.20	—	—	—	—
1030	0.28-0.34	0.60-0.90	0.10-0.20	—	—	—	—
1035	0.32-0.38	0.60-0.90	0.10-0.20	—	—	—	—
1040	0.37-0.44	0.60-0.90	0.10-0.20	—	—	—	—
4023	0.20-0.25	0.70-0.90	0.20-0.35	—	—	0.20-0.30	—
4130	0.28-0.33	0.40-0.60	0.20-0.35	—	0.80-1.10	0.15-0.25	—
4140	0.38-0.43	0.75-1.00	0.20-0.35	—	0.80-1.10	0.15-0.25	—
4615	0.13-0.18	0.45-0.65	0.20-0.35	1.65-2.00	—	0.20-0.30	—
5015	0.12-0.17	0.30-0.50	0.20-0.35	—	0.30-0.50	—	—
5120	0.17/0.22	0.70-0.90	0.20/0.35	—	0.70-0.90	—	—
5130	0.28-0.33	0.70-0.90	0.20-0.35	—	0.80-1.10	—	—
6120	0.17-0.22	0.70-0.90	0.20-0.35	—	0.70-0.90	—	0.10-0.15
8620	0.18-0.23	0.70-0.90	0.20-0.35	0.40-0.70	0.40-0.60	0.15-0.25	—
8630	0.28-0.33	0.70-0.90	0.20-0.35	0.40-0.70	0.40-0.60	0.15-0.25	—
51410	0.15 max	—	—	—	11.5-13.5	—	—

60 per cent and to employ 40 per cent reductions for subsequent stages. Cold extrusion of steel is frequently used also as a stage in the production of parts by deep drawing. Another procedure is to warm either the blanks or the tools; but this procedure requires careful selection of the temperature to avoid entering the blue brittle range.

Melting Practice

Although rimmed steels are sometimes cold-extruded, in general the aluminum-killed, fine-grain steels are preferred. According to the AISI,[4] steel for cold-extruded parts is produced with closely controlled practice and there is rigid testing to assure internal soundness and freedom from surface imperfections.

An excess of aluminum in aluminum-killed steels is used to prevent or restrict strain aging, which is a valuable consideration in cold extrusion as in other cold-forming operations. Hauttmann and Pessl[5] made an extensive investigation in which they compared the characteristics of cold-extruded aluminum-killed steels carrying excess aluminum (0.025 to 0.05 per cent) with those of steels that contained no aluminum. Carbon contents of these steels ranged from 0.08 to 0.28 per cent. They reported that no difficulties were encountered in the cold extrusion of the aluminum-bearing steels, provided they were properly heat-treated; the desired ductility was attained by normalizing, followed by relatively rapid cooling. As a result of their work, they concluded that the aluminum-bearing steels retained a higher degree of ductility than steels of similar composition not containing aluminum.

Structure

As mentioned previously, the structure of the steel is an important factor in extrudability. In general, a soft material is desirable to minimize the extrusion pressure. Optimum results are attained if the hardness of the blank does not exceed Rockwell B60 to B80. For a given composition, a structure containing fairly coarse spheroidized carbide should require the lowest extrusion pressure. Such a structure, however, is generally economically impractical because of the long time the steel must be held in the subcritical range.

Howard, Dennison, and Angus[6] investigated the effects of

various heat-treatments on a series of steels. These treatments included full annealing at a number of temperatures, normalizing, and quenching and tempering. They concluded that the lowest extrusion pressures were attained on the fully annealed material with little variation resulting from differences in annealing temperatures. Normalized material had extrusion pressures about three to five per cent greater than the fully annealed material; quenched and tempered specimens required considerably higher pressures. They also determined that hot-rolled material, as received from the mill, required about the same extrusion pressures as the normalized materials, and concluded that hot-rolled material should be suitable for numerous applications.

Okell[2] states that, usually, cooling from the subcritical range after a reasonable soaking time is all that is required for low-carbon steels and that little is gained by attempting to spheroidize the carbides in simple low-carbon steels. With alloy steels, however, punch loads are higher and it is desirable to place as many as possible of the alloying constituents in the carbide phase, in order to reduce the quantities of alloying elements in the ferrite, and to develop a continuous ferrite phase. This objective is achieved by cooling as slowly as possible to allow maximum precipitation of the alloying constituents in the ferrite.

Morgan[7] goes into a little more detail. He states that annealing at about 1300 F is satisfactory for plain carbon steels containing not more than 0.20 per cent of carbon, and that full spheroidization is not necessary. As an example, he notes that holding a billet $1\frac{1}{4}$ inches in diameter for 45 minutes at 1300 F is sufficient for the annealing of these low carbon steels. With carbon contents above 0.30 per cent, full spheroidization is necessary and annealing times up to four hours at 1300 F may be required. Annealing should be followed by cooling in the furnace, if possible. He states that a 0.20%-carbon steel with a Vickers hardness of 120, or a 0.30%-carbon steel with a hardness of 135, is suitable for extrusion.

One procedure that has been used successfully for commercial operations differs from these suggestions. This method requires: 1) Soaking above the critical range for sufficient time to attain uniformity; 2) quenching in water to retain uniform distribution of the carbides; and 3) annealing at a suitable temperature to produce rapid softening but retain a fine granular structure. This treatment is reported to result in optimum distribution of carbide particles and softening.

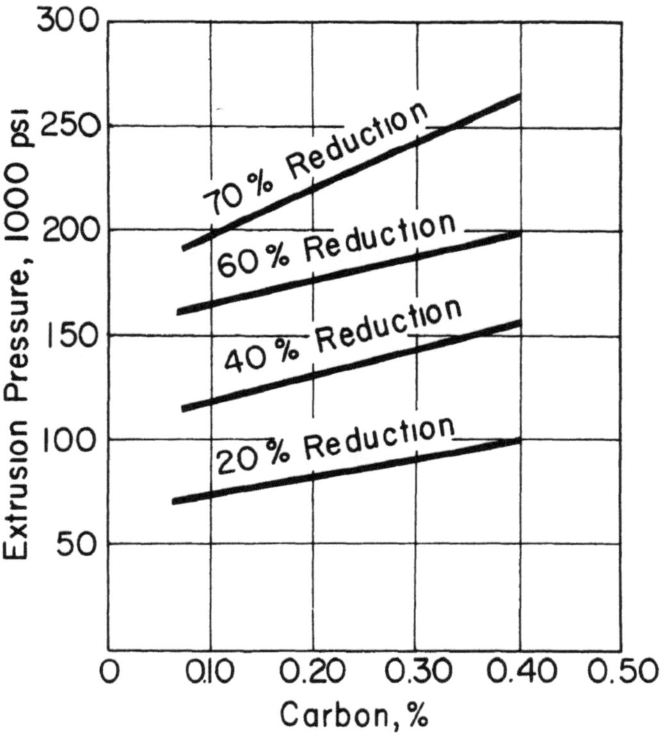

FIG. 12:1. EFFECT OF CARBON CONTENT ON THE EXTRUSION PRESSURE IN FORWARD EXTRUSION (Jonck[8])

TABLE 12:2

COMPOSITIONS AND HEAT TREATMENTS OF STEELS USED IN PLOTTING FIGURE 12:1

Composition, %					*Brinell Hardness*
C	Si	Mn	P	S	
0.07	0.08	0.27	0.015	0.025	104
0.12	0.27	0.33	0.012	0.028	109
0.20	0.22	0.43	0.022	0.036	122
0.40	0.29	0.59	0.024	0.018	155

All steels heat-treated as follows: 1640 F, 20 min, air-cooled; 1300 F, 4 hr, air-cooled.

EXTRUSION PRESSURES

The extrusion pressures required are affected by the composition of the steel and its mechanical properties. For example, Jonck[8] states that the extrusion pressure for a given reduction of area is a linear function of the carbon content in plain carbon steels. His data are plotted in Figure 12:1 for the steels, and the treatments given are in Table 12:2.

The extrusion pressure depends also to a great degree on the surface treatment—it was impractical to cold-extrude steel until the phosphate treatment was devised (the subject of lubrication is discussed in Chapter 6). A number of investigators have reported extrusion pressures for various steels but the results are comparable only within their own groups because of differences in lubrication procedures. Some of these results, however, will be used to indicate the pressures involved.

The Heintz Manufacturing Company made an extensive investigation of the cold extrusion of a number of steels.[9] They

TABLE 12:3

EXTRUSION LOADS FOR VARIOUS REDUCTIONS OF AREA*

Material	*Rockwell B Hardness*[a]*	*Reduction of Area %*	*Starting Extrusion Load, lb*	*Final Extrusion Load, lb*
Ingot Iron	29	40	40,000	30,000
		60	83,000	72,000
		70	139,000	116,000
C1010 Rimmed	45	40	57,000	40,000
		60	106,000	73,000
		70	150,000	122,000
C1010 Killed	58	40	53,000	40,000
		60	108,000	83,000
		70	166,000	132,000
C1020	57	400	59,000	42,000
		60	118,000	89,000
		70	171,000	144,000
C1040	81	40	63,000	53,000
A4615	72	40	62,000	48,000
		60	145,000	104,000
A5120	74	40	66,000	82,000
		60	141,000	110,000

[a] Before extrusion
* Heintz Manufacturing Co[9]

TABLE 12:4

COMPOSITIONS AND HEAT-TREATMENTS OF STEELS LISTED IN TABLE 12:3

Material	*Composition %*							*Heat-Treating Temperature F*	
	C	Mn	Si	Al	Ni	Cr	Mo	Normalizing	Annealing
Ingot Iron	0.08	0.05	0.022	0.003	0.056	0.02	0.02	1675	1550
C1010, Rimmed	0.08	0.37	0.002	0.005	—	—	—	1675	1550
C1010, Al-Killed	0.12	0.50	0.031	0.031	—	—	—	1675	1550
C1020, Al-Killed	0.21	0.48	0.005	—	—	—	—	1675	1550
C1040, Si-Killed	0.38	0.80	0.027	0.027	—	—	—	1650	1450
A4615	0.16	0.47	—	—	1.78	0.11	0.23	1750	1600
A5120	0.20	0.78	—	—	—	0.80	—	1650	1475

* Heintz Manufacturing Co.[9]

TABLE 12:5

EXTRUSION PRESSURE FOR THE FORWARD EXTRUSION OF VARIOUS STEELS*

Key[a] *No.*	*Composition %*				*Annealed Tensile Strength psi*	*Extrusion Pressure (psi) for Reduction of Area of*			
	C	Mn	Si	Al		30%	40%	50%	60%
A	0.08	0.33	0.07	0.059	50,000	57,000	85,000	99,000	150,000
B	0.10	0.41	—	—	51,000	57,000	85,000	128,000	142,000
C	0.09	0.53	0.10	0.025	52,000	57,000	81,000	128,000	170,000
D	0.11	0.45	0.10	—	54,000	57,000	85,000	114,000	170,000
E	0.16	0.48	0.12	—	63,000	71,000	99,000	157,000	193,000
F	0.24	0.50	0.28	0.040	71,000	71,000	92,000	142,000	200,000
G	0.45	0.80	0.24	—	97,000	85,000	128,000	193,000	280,000
H	0.50	0.90	0.20	—	111,000	114,000	142,000	200,000	295,000

[a] A, Special Steel for Cold Extrusion
B, Rimmed Steel for Deep Drawing
C, Nonaging Boiler Plate Steel
D, Killed Boiler Plate Steel
E, Carburizing Steel
F, Nonaging Boiler Plate Steel
G, Heat Treatable Structural Steel
H, Thomas Converter Rail Steel

* Hauttmann and Pessl[5]

reported their extrusion pressures in total load rather than in load for unit of area. Some of their results are given in Table 12:3. Compositions and heat-treatments of these steels are listed in Table 12:4.

In discussing these results, Bowden[10] stated that at a reduction of area of 20 per cent using a steel containing not more than 0.20 per cent carbon, the working stress on the extrusion shoulder in forward extrusion was about 115,000 *psi*. At 40 per cent reduction of area, the stress ranged from 225,000 to 275,000 *psi*. In their work, lower stresses were encountered at 60 to 70 per cent reduction of area than at 40 per cent; at 75 per cent reduction, the stress was the same as that at 40 per cent reduction.

Hauttmann and Pessl[5] reported the results obtained during the extrusion of a number of German steels. Their data, given in Table 12:5, indicate that the pressure increases with the carbon content, as noted by Jonck. There appears to be no definite correlation, however, between melting practice and extrusion pressure although they state that the aluminum-bearing steels are preferred to the aluminum-free steels.

TABLE 12:6

EXTRUSION PRESSURES FOR BACKWARD EXTRUSION*

Composition %					*Extrusion Pressure for 50% Reduction of Area psi*
C	Mn	Si	Ni	Cr	
0.10	0.70	0.18	—	—	294,000
0.27	0.33	0.48	0.20	13.2	425,000
0.03	0.60	0.73	9.92	18.5	505,000
0.03	1.04	0.89	12.8	13.0	440,000

* Ridings and Angus[11]

Ridings and Angus[11] investigated the extrusion pressures required for the backward extrusion of a number of materials including several steels. Their data are listed in Table 12:6. Although these were laboratory experiments, the data are interesting in showing that the stainless steels can be backward-extruded with significant reductions of area, although the extrusion pressures are quite high.

During the course of a survey on the process, a number of producers noted that for plain carbon steels, a good estimate of the required pressure could be obtained from the tensile strength. As a rough approximation, the extrusion pressure for 65 per cent reduction of area can be taken as three times the tensile strength of a normalized low-carbon steel.

MECHANICAL PROPERTIES OF EXTRUDED PARTS

One of the most interesting features of cold extrusion of steel is its influence on the mechanical properties of extruded parts. The extensive cold-working roughly doubles the hardness and

TABLE 12:7
IMPROVEMENT IN MECHANICAL PROPERTIES RESULTING FROM COLD EXTRUSION*

Material	*Annealed Properties*			*Cold Reduction %*	*Worked Properties*		
	Tensile Strength psi	Yield Strength[a] psi	Elongation (2 in.)%		Tensile Strength psi	Yield Strength[a] psi	Elongation (2 in.)%
Ingot Iron	42,000	18,000	45	40	86,000	84,000	13
				70	104,000	103,000	13
C1010 (Rimmed)	46,000	29,000	45	40	94,000	92,000	16
				70	108,000	107,000	12
C1010 (Killed)	52,000	36,000	44	40	91,000	85,000	16
				70	107,000	101,000	14
MBK6[b]	51,000	32,000[c]	30	—	82,000	57,000[d]	10
CK10[b]	54,000	36,000[c]	25	—	86,000	57,000[d]	10
CK15[b]	58,000	40,000[c]	20	—	93,000	71,000[d]	8
1018	60,000	29,000	39	[e]	146,000	134,000	12
C1020	58,000	40,000	38	40	105,000	101,000	9
				70	123,000	122,000	11
CK35[b]	66,000	44,000[c]	20	—	103,000	86,000[d]	6
C1040	75,000	50,000	35	40	126,000	113,000	12
4130	81,000	52,000	28	[e]	166,000	139,000	14
4140	95,000	60,000	26	[e]	178,000	159,000	12
A4615	71,000	46,000	34	40	111,000	100,000	8
A5120	77,000	48,000	32	40	116,000	104,000	16
				60	129,000	120,000	16
8620	78,000	56,000	31	[e]	144,000	126,000	14
8630	82,000	54,000	29	[e]	171,000	166,000	12
15Cr3[b]	61,000	43,000[c]	20	—	93,000	71,000[d]	12
16MnCr5[b]	66,000	48,000[c]	20	—	93,000	71,000[d]	12

[a] 0.2% offset
[b] German Steels

	Composition %			
	C	Mn	Si	Cr
MbK6 (Al-killed)	0.10	0.30	0.10	—
CK10 (Si-killed)	0.10	0.35	0.25	—
CK15 (Si-killed)	0.15	0.35	0.25	—
CK35 (Si-killed)	0.35	0.50	0.35	—
15Cr3	0.15	0.50	0.25	0.65
16MnCr5	0.17	1.20	0.25	0.90

[c] Proof stress, maximum
[d] Proof stress, minimum
[e] Six forming operations including backward and forward extrusion
* Heintz,[9] Feldmann,[12] King[13]

tensile strength of low-carbon steels. The effect is even more pronounced in the improvement of yield strength, which may be raised by a factor of three. The actual values obtained on low-carbon steels, which may have yield strengths of the order of 100,000 *psi*, approach those obtainable by the heat-treatment of low-alloy steels. As a result, several applications have been developed in which parts made of cold-extruded carbon steel have displaced parts made of heat-treated alloy steel with considerable reduction in costs.

Data on ingot iron and a number of steels, included in Table 12:7, confirm the improvements resulting from cold extrusion. For example, a reduction of area of 40 per cent increased the yield strength of ingot iron from 18,000 to 84,000 *psi*, and a 70 per cent reduction increased it to 103,000 *psi*. At the same time, considerable ductility, as measured by the elongation, remained. Similar results were obtained on the various steels listed in the table although the improvement in properties was not always so great. An examination of the table indicates that:

At a 40 per cent reduction of area, the yield strength was increased as much as 500 per cent and the tensile strength up to 200 per cent.

At 70 per cent reduction of area, the yield strength was increased as much as 600 per cent and the tensile strength as much as 250 per cent.

It is apparent from the table that the low-carbon steels show the greatest improvement in properties. As these steels also require the lowest extrusion pressures, it is advantageous to use them for as many applications as possible.

SIZES AND TOLERANCES ON PARTS IN COMMERCIAL PRODUCTION

The cross-sectional area and length of a part that can be produced by cold extrusion depend on the equipment available, but they are also influenced by limitations in the process. In backward extrusion, the strength of the column of the punch is a limiting factor. The length of a steel shell is generally limited to three or four times the diameter. Parts up to six inches in diameter and 20 inches long are being produced commercially. If the ratio of length to diameter is maintained, the size of the part is limited only by the capacity of the press. As larger presses are being developed, it can be expected that larger parts than those mentioned will be available in the near future.

In forward extrusion, the punch can be more sturdy than that used in backward extrusion and strength of the column is not a limiting factor. Tubes up to 14 inches in diameter and 10 feet long have been produced.

A major advantage of cold extrusion is the ability to hold close tolerances on the thickness of wall and bottom, and on outside and inside diameters. The length shows greater variation unless the size of the blank is very carefully controlled. Frequently, the tolerance for length depends on a secondary operation—trimming the shell. The cold extrusion of steel has not been developed to the same degree as the impact extrusion of aluminum and there appear to be no standard tolerances. Some idea of those obtainable can be determined from the data published by Feldmann.[1] He suggests a tolerance on the as-extruded length of 0.12 to 0.24 inch; otherwise the part must be trimmed to length. Acceptable deviation of the outside diameter from the inside diameter should be 0.5 to 1.2 per cent of the nominal outside diameter. Out-of-roundness should be 0.2 to 0.6 per cent of the nominal outside diameter.

On the forward extrusion of solid parts, he states that the diameter can be held to ± 0.002 to ± 0.5 inch depending on the size of the part. On backward-extruded parts, tolerances for the outside diameter can be held to ± 0.001 inch for small parts and to ± 0.016 inch for large parts; inside diameter to ± 0.004 inch for small parts and ± 0.014 inch for large ones.

Compared with some of the tolerances given in some of the examples listed later, the recommendations of Feldmann appear to be somewhat conservative. The tolerances, however, are sufficiently close to permit the use of many parts without additional sizing operations, such as ironing or coining to increase the dimensional accuracy.

The surface finish obtainable on the as-extruded part depends on the finish maintained on the tools. Surface finishes as smooth as 30 microinch *rms* are obtainable and those of 60 to 200 microinch *rms* are normal. The finished surface, however, depends on whether the residual phosphate coating is permitted to remain or must be removed to permit joining operations. If the coating must be removed, some of the surface finish must be sacrificed because the removal of phosphate roughens the surface. If the finish must be removed sufficient stock must be allowed on the part to permit a machining operation.

APPLICATIONS

The cold extrusion of steel is being extended to an ever-increasing number of fields as its advantages become more widely recognized. When the shape of the part permits use of cold extrusion, either as a direct substitution for another manufacturing method or with a change in design, the process often results in considerable reduction in costs of material and machining.

In the nonferrous field, most parts are produced by a single extrusion operation which may be followed by spinning, bulging, and various machining operations. In the ferrous field, cold extrusion is frequently performed in several steps which may include backward or forward extrusion, or a combination of the two. Cold extrusion is also used as a step in a deep-drawing operation.

The advancement of cold extrusion is indicated by the increase in the quantity of steel being used for the purpose. During the five years ending in 1962, shipments of steel for cold extrusion increased from 50,000 tons to 200,000 tons.[14] Seventy per cent of this steel goes to the automotive industry, and SAE 1008 is the most widely used steel.

Some specific examples of parts produced by cold extrusion are given in the following sections. As the automotive industry has made the greatest advances in the application of the process to steel, the major portion of the following discussion is devoted to automotive parts.

Automotive Industry

As the automotive industry is one of the major mass-production industries, it is natural that a rapid method which offers the possibility of saving material and labor should be widely utilized in this field. Some of the parts that are in production are:

Truck Wheel Nut: A nut for the wheel of a truck was in production as a screw-machine part and 18 ounces of stock was required. The weight of raw material was reduced to 12 ounces by processing on a screw-machine and a header. Converted to a cold-extruded part, the weight of stock was reduced to 6 ounces and the completed part weighed $5\frac{3}{4}$ ounces. The only machining required on the extruded part is tapping the inside threads and rolling the outside threads. This part is produced at a rate of 50 pieces a minute.

Torsion Bar Anchor: The suspension control arm bushing was produced from a bar of stock SAE 1035 hot-rolled by machining and boring on a screw-machine, according to Garwood.[15] The hexagonal hole in one end was broached and the part required heat-treatment to meet the specifications for mechanical properties. The part is now produced from a bar of SAE 1021, spherodize-annealed stock. The blank, $2\frac{3}{32}$ inches in diameter

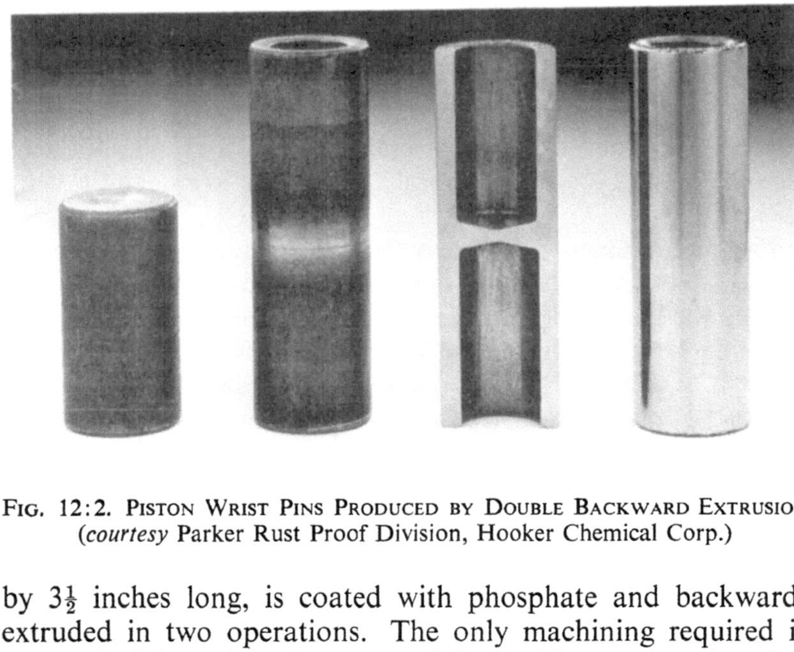

FIG. 12:2. PISTON WRIST PINS PRODUCED BY DOUBLE BACKWARD EXTRUSION (*courtesy* Parker Rust Proof Division, Hooker Chemical Corp.)

by $3\frac{1}{2}$ inches long, is coated with phosphate and backward-extruded in two operations. The only machining required is removal of the web in the center of the bushing, and cutting ring grooves on one end. The saving in cost for each part amounts to seventeen cents.

Spark Plug Shell: The shell for a spark plug, formerly produced on an automatic screw-machine, was converted to a cold extrusion with an over-all saving of 65 per cent of steel and 86 per cent reduction in time of production.

Hydraulic Steering Worm: This worm gear was originally machined from bars of SAE 4028. Garwood[15] states that the part is now being forward-extruded from SAE 4027 bars, the conversion being made only to save machining time. All other operations, which include threading, heat treatment, and cutting grooves, remain unchanged.

Universal Joint Bearing Cups: The bearing cups, used on propeller shaft assemblies, were formerly produced on automatic screw-machines.[16] The cups are now being produced from SAE 1012 hot-rolled steel coil stock. The stock is cold-drawn to the desired size and fed into a machine that cuts the blank, squares it, and chamfers both sides. Blanks are then washed, blasted with grit, annealed, and phosphate-coated in a continuous operation. Two bearing cups are extruded at one time. They are finished by conventional procedures. Finished tolerances are 1.158 to 1.160 inches, outside diameter; and 0.843 to 0.845 inch, inside diameter. The parts are produced at the rate of 3000 an hour.

Wrist Pins: Piston wrist pins are produced conventionally by drilling bar stock on automatic machines, the operation requiring about three-fourths of a pound of stock for each pin, of which about one-fourth of a pound is scrap.[17] These pins are now being cold-extruded from SAE 5015 fine-grain, cold-heading bars. The bars are cut into chamfered blanks on automatic machines, phosphate-coated, and lubricated. The pins are formed by double backward extrusion, the punches entering both ends of the blank simultaneously. The short solid web that remains at the center is removed by machining. Production rates are 2000 pieces an hour (see Figure 12:2).

Differential Gear for Truck: According to Farr,[17] a differential gear for a truck, formerly hot-forged and machined by broaching the splines and hobbing the gear teeth, has been converted to a cold extruded part. Starting with a 7.6-pound billet of SAE 8620 steel, both splines and teeth are extruded to the tolerances required. The only machining necessary is facing both ends and some undercutting and relieving. The cost of production was reduced 20 per cent by converting this part to cold extrusion.

Needle Bearing Races for Universal Joint: These races are produced from strips of SAE 1010, hot-rolled, pickled, and oiled. The slugs are blanked eleven at a time and forward-extruded into cups, using a magazine feed of 35 pieces a minute.

Propeller Shaft: A propeller shaft formerly produced from SAE 1045 steel by forging was converted to a cold-extrusion part using SAE 1035. The saving in material was about one-fifth of a pound for each part.[18] The item is forward-extruded to form the shaft, and backward-extruded to form the hexagonal socket. The rate of production is 720 to 750 parts an hour.

Switch for Stop Lamp: The housing for the switch of a stop lamp is produced from SAE 1008 hot rolled rimmed steel rod

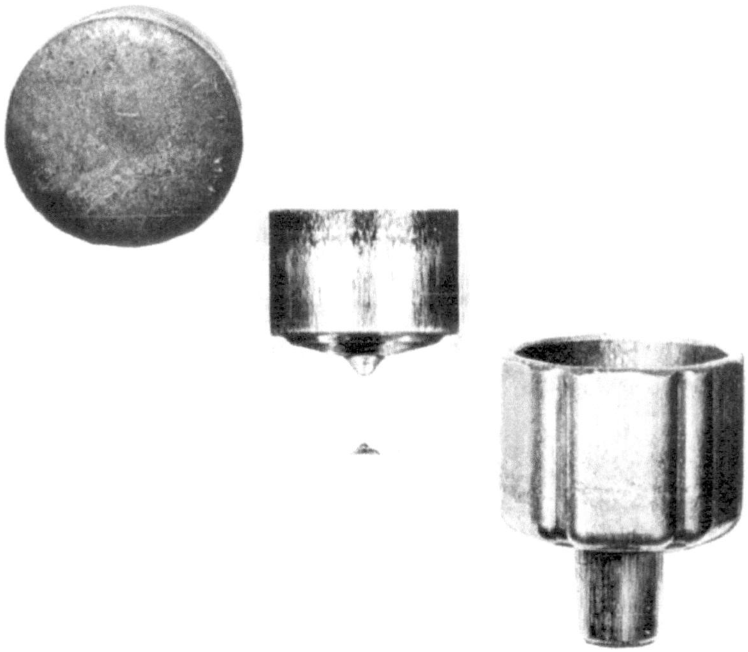

FIG. 12:3. HOUSING FOR SWITCH OF STOP LIGHT EXTRUDED FROM CARBON STEEL (*courtesy* Braun Engineering Company)

which is $\frac{5}{8}$-inch in diameter. Suitable lengths are cut off, phosphate-coated, lubricated, and upset to a blank one inch in diameter, at the rate of 160 a minute. The blanks are degreased, cleaned, annealed, phosphate-coated, lubricated, and extruded. The housing is completed by machining and threading (see Figure 12:3).

Housing for Rear Axle of Truck: This housing is being produced by cold extrusion from SAE 1030 steel at a reduction of 17 per cent in material over the method used previously.[20] Slugs $5\frac{3}{8}$ inches in diameter and three inches long are sawed from a billet, annealed, phosphate-coated, and lubricated. They are backward-extruded to form the center hole, annealed at 1200 F, recoated and lubricated, and forward-extruded. They are re-annealed at 1200 F, recoated and lubricated, and again forward-extruded. The next operation is ironing to reduce the thickness of the wall on the neck of the spindle. The rough finished part is 12 inches long with a wall $\frac{1}{8}$ inch thick at the neck. The ends of the neck are trimmed and the spindles are welded to the ends of the housing of the axle. Finally, the welded assembly is water-quenched from 1800 F, and tempered to the desired hardness. The production line is to be automated and it is expected to produce parts at the rate of 350 an hour.

Miscellaneous: Among other automotive parts being produced by cold extrusion are:

A one-piece power steering cylinder for trucks, which has supplemented a welded assembly made from a forging and a seamless tube.

A torsion tube for truck tandem axles in place of a welded assembly.

A four-stage cold extrusion operation is being used to produce automobile ball-joint, cup-shaped housings.

In addition, transmission shafts, bolts, nuts, valve lifters, brake-adjusting clevis pins, front-end ball suspensions, and engine tappets are being produced commercially.

In experimental production are bodies for complex spark plugs, couplings for carburettor fuel lines, fittings, and many other parts.

General Industry

An indication of the variety of parts being cold-extruded at present can be obtained from the following examples.

Socket Wrenches: Socket wrenches are being cold-extruded within tolerances of ± 0.002 inch.[21] The raw material, a bar of SAE 50B44 steel one inch in diameter, is coined to a blank, 1.23 inch in diameter which is then spheroidized by annealing. After the annealing, the blank is sand-blasted, phosphate-coated, and lubricated. Double backward extrusion is used to form the opening in the hexagonal nut and the half-inch-square hole for the ratchet-driver. The only machining required is facing and chamfering the holes, and cutting grooves in the external surface of the socket to permit gripping for removal from the handle. The sockets are hardened by heat-treatment, polished, and plated with nickel-chromium. Savings in material over the previous method are expected to range from 40 to 60 per cent, depending on the size of the socket.

Accumulator Shell: An accumulator shell for the hydraulic cranking system in engines used in the oil and gas fields is being cold-extruded in five steps. The shells range from 16 to 42 inches in length and are $4\frac{3}{4}$ inches in diameter. Tolerances of ± 0.0025 to ± 0.004 inch are held in the sections of the shell within which the piston travels, and out-of-roundness is held within ± 0.002 inch. Production involves both backward and forward extrusion. The cold-extruded part displaces a conventionally fabricated assembly of a seamless tube and a forging, which had required machining, threading, and the use of an O-ring seal. The cold-extruded part is in one piece, requiring no O-ring seal.

Outboard Motor Propeller Shaft: A propeller shaft for an outboard motor is being cold-extruded from bar stock of Type 410 chromium stainless steel, heat-treated to the range of Rockwell C36 to C40.[22,23] As the usual zinc phosphate treatments do not produce an adherent coating on stainless steel, the surface is first treated with an oxalate coating and a complex soap-type lubricant is then applied. Stepped shafts are cold-extruded in five diameters, the shaft being nominally one inch in diameter and $12\frac{3}{16}$ inches long, with major diameters held to ± 0.005 inch tolerances. Cold extrusion eliminates four machining operations and reduces scrap substantially. The part is completed by grinding the surface to the desired finish, forming a spline by rolling, cutting a key-slot, and drilling a hole for a cotter-pin.

Pressure Tank: In a tank for an oxygen pressure system, a cold-extruded part supplanted a hot-spun part, primarily because it was possible to improve the concentricity of the outside and inside diameters. With cold extrusion, it was possible to hold the concentricity to ± 0.020 inch.

Among other parts that have been produced by cold extrusion are pistons for compressors, hollow gears, housings for generators, and rollers for crawler tractors; the last are carburized before use. A thrust plate produced from stainless steel is shown in Figure 12:4.

Ordnance

The application of cold extrusion to the production of ordnance components has been quite successful. For example, Farr[17] mentions that production of the 105-mm shell has reduced raw material requirements from 42 pounds for a forged shell to 27 pounds for a cold-extruded shell, and reduced the tolerance in weight from 9 to 2 ounces. Among other shells produced, according to Farr, are 20-mm, 75-mm, 76-mm, 105-mm, and the 2.75-inch rocket.

FIG. 12:4. THRUST PLATE PRODUCED FROM TYPE 410 STAINLESS STEEL (*courtesy* Steel Industries, Inc.)

Rocket Head: Cold extrusion is one stage in producing the head of a 2.75-inch rocket. According to Turner,[24] the steps in production of this part are sizing the blank, cupping by backward extrusion, drawing, coining the base, and " nosing " the open end.

The blank, which is $3\frac{1}{8}$ inches in diamter by $2\frac{1}{2}$ inches long and weighs $5\frac{1}{2}$ pounds, is cut from a hot-rolled bar of aluminum-silicon-killed 1010 steel. Cold-working during sizing and backward extrusion raises the hardness from about Rockwell B60 to B94. Before further processing, the cup is annealed at 1500 F to reduce the hardness to Rockwell B60-65. The annealed cup is drawn, reannealed at 1450 F, coined, and nosed.

75-mm Shell: A 75-mm shell is now in production by cold-extrusion from C1018 fully aluminum-killed steel. After being sized, the formed blank is annealed at 1200 F, phosphate-coated and lubricated, and backward-extruded. Next, it is reannealed, recoated, and relubricated. Then it is forward-extruded, coined, drawn, and nosed. The finished shell is stress-relieved at 800 F. Parts are produced at the rate of 250 an hour. This procedure saves considerable material and labor over the method of hot-forging which is also used in the production of this shell.

81-mm Mortar Shell: Backward extrusion is combined with conventional deep drawing to produce the 81-mm mortar shell. After being sized, coated, and lubricated, the part is formed by backward extrusion, with a reduction of area of 54 per cent. The remaining steps are conventional redrawings. Backward extrusion eliminates one redrawing step.

8-inch Shell: According to Babcock,[25] the 8-inch shell can be produced by a process that combines hot cupping, machining, and cold extrusion. The slug is cupped, machined to the proper size, phosphate-coated, lubricated, and forward-extruded. The remaining steps are coining the base, cold-drawing, expanding the bourrelet, trimming to length, beveling the taper, and cold nosing. This process reduces the quantity of steel used in each shell by 30 per cent over the conventional method of piercing and drawing.

REFERENCES

1. FELDMANN, H. D., Economical Use of Cold-Forged Components, *Sheet Metal Industries*, 38, January 1961:8.
2. OKELL, R. E., Metallurgical Aspects of the Cold Extrusion of Steel, *Metal Treatment*, 28:309 (1961).
3. VERSON, M. D., Pressure Economy, *Sheet Metal Industries*, 38, July 1961:493.
4. Carbon Steel; Semifinished for Forging; Hot Rolled and Cold Finished Bars; Hot Rolled Deformed Reinforcing Bars, *Steel Products Manual*, AISI, New York, 1957.
5. HAUTTMANN, H., and PESSL, H. J., The Aluminum-Bearing, Non-Aging Steels and Their Cold Extrusion, *Steel Processing*, November 1950:557.

6. HOWARD, F., DENNISON, H. A. J., and ANGUS, N., Some Investigations Into the Cold Extrusion of Steel, *Sheet Metal Industries*, 38, June 1961:403.
7. MORGAN, R. A. P., The Cold Extrusion of Steel, *Jour. Iron and Steel Inst.*, 193:285 (1959).
8. JONCK, R., The Influence of Composition and Structure on the Resistance of Steels to Deformation in the Forming of Solid Parts, *Revue de Metallurgie*, 59, January 1962:41.
9. *Report of Metallurgical Research on Cold Extrusion of Steel*, Heintz Manufacturing Company, September 1948 (OTS, PB 135 108).
10. BOWDEN, J. P., Cold Extrusion of Steel, *Steel Processing*, May 1950:227.
11. RIDINGS, J. B. and ANGUS, N. S., Experiments on the Reverse Impact Extrusion of Certain Metals, *Jour. Inst. Metals*, 90, December 1961:107.
12. FELDMANN, H. D., Properties and Possibilities of the Use of Cold Extruded Steel Parts, *Memoires Soc. de Ingenieurs de France*, July-August 1956:247.
13. KING, J. E., American Acceptance of Heavy Cold Forming of Steel, ASME Diamond Jubilee Meeting, Baltimore, April 19, 1955.
14. Cold Forming Winning Favor in Detroit, *Steel* March, 26, 1962: 83.
15. GARWOOD, M. F., Materials and Process Applications in the Automotive Industry, *Metal Progress*, October 1960:115.
16. Bearing Cups Cold Formed, *Steel*, February 23, 1959:93.
17. FARR, C. P., Extrusion Method Intriguing, *Canadian Metalworking*, 25, February, 1962:26.
18. Extrusion Helps Hold Cost Line, *Steel*, August 31, 1959:64.
19. BRAUN, A., Cold Extrusion of Steel Parts from Wire, *Wire and Wire Products*, 35:1538 (1960).
20. AMP Cold Extrudes King Size Part, *Steel*, February 5, 1962:51.
21. Cold Impact Alters Grain Flow to Boost Socket Strength, *Iron Age*, June 22, 1961:130.
22. ESHELMAN, R. H., Cold Extrude Stainless Shafts, *Ibid.*, January 21, 1960:86.
23. For Outboard Motors; Stainless Shafts Cold Extruded, *J & L Explorer*, 1, No. 5:9 (1960).
24. TURNER, C. A., Jr., Short Cycle Anneal Restores Ductility in Cold Extrusions, *Iron Age*, March 10, 1955:96.
25. BABCOCK, H. J., Process Engineering from Prototype to Production of the 8-inch M106 H.E. Shell by the Hot Cup—Machine—Cold Extrude Methods, *Proc. Industrial Tech. Conf. on Metal Processing*, October 1957 (OTS, PB 151 308).

Selected Bibliography

Arnold, F. M., How to Pick the Right Forming Lubricant, *Steel*, November 21, 1960:98.

ASM Committee, The Selection of Material for Cold Extrusion Tools, *Metals Handbook*, *Vol. 1*, 8th Edn.; American Society of Metals, 1961:720.

Babcock, H. J., Process Engineering from Prototype to Production of the 8-inch M106 H.E. Shell by the Hot Cup—Machine—Cold Extrude Method, *Proc. Ind. Tech. Conf. on Metals Processing*, October 16-18, 1951, PB 151308 (Office Technical Services, U.S. Department of Commerce).

Balduzzi, P., The Manufacture by Impact Extrusion of Aluminum Cylindrical Containers with Flanged Tops, *Alluminio*, 20:553 (1951).

Boston, J. R., Cold Extrusion of Aluminum Parts for Douglas Aircraft, *Machinery*, 51, July 1945:138.

Butler, L. H., Surface Conformation of Metals Under High Nominal Contact Pressures, *Metallurgia*, 61, April 1960:167.

—, The Effect of Lubricants on the Growth of Surface Contact Areas during Plastic Deformation of Metals, *Jour. Inst. Metals*, 89:116 (1960).

Byles, H. W., Impact Extrusion, *Metal Treatment*, 25, March 1958: 107.

Chopin, R. and Chopin, J., Impact Extrusion of Light Metals, *Revue de L'Aluminium*, 32 December 1955:1150.

Chopin, R., Researches on the Technique of Impact Extrusion, *Ibid.*, 33, April 1956:365.

Comley, A. W. F., Tool Materials for the Cold Extrusion Process, *Sheet Metal Industries*, 38, March 1961:190.

Cooper, A. M., Basic Features of the Cold Forging Process, *Wire Industry*, 27, December 1960:1189.

Dallas, D. B., Principles of Cold Forming: Part 1, Designing the Part, *Tool and Manufacturing Engineer*, 47, November 1961:101.

Davis, D. J., *Cold Working of Metals*, Society Automotive Engineers, Preprint S65, February 1958.

Dipper, M., Calculating and Determining the Pressures Required for the Cold Extrusion of Shells, *Archiv. Eisenhüttenwesen*, 20:275 (1949).

Durer, A., Phosphate Coating as an Aid to the Shaping of Steel, *Ibid.*, 20:305 (1949).

EBERSPAECHER, H., Cold Extrusion of Steel, *Werkstatt und Betrieb*, 91:605 (1958).
EVERHART, J. L., Impact (Cold) Extruded Parts, *Materials & Methods*, August 1955:111.
FELDMANN, H. D., Investigation on the Power and Work Required in Cold Forging of Various Steels, *Stahl und Eisen*, 73:165 (1953).
—, Industrial Application of Cold-Flow Extruded Steel Parts, *Draht* (English Edn.) February 1957:18.
—, Design Principles for Cold Pressed Steel Components, *Konstruktion*, 3:82 (1959).
—, The Design and Use of Cold-Forged Components, *Sheet Metal Industries*, 37:325 (1960).
—, Press Selection for Cold Forging, *Metalworking Production*, 105, January 11, 1961:51.
—, Economical Use of Cold Forged Components, *Sheet Metal Industries*, 38, January 1961:8.
—, Production Examples for Cold Forging, *Draht*, 12:637 (1961).
—, The Effect of Cold Forging on the Quality of Steel Components, *Research Applied in Industry*, 14, December 1961:487.
—, Cold Forging of Steel (translated into English by A. M. Hayward). Chemical Publishing Company, New York, 1962.
—, A Comparison of Hot and Cold Forging of Steel, *Metal Treatment*, January 1962:13.
FIORENTINO, R. J., SABROFF, A. M., and Frost, P. D., Development of Improved Methods for Cold Extrusion of Titanium, Air Material Command, Technical Report 61-557, 1961.
FOISY, G. A., The Manufacture and Application of Extruded Copper Tubes, *Trans. Amer. Soc. Mech. Engineers*, 1927, MSP 50.
Forward Extrusion of Steel Reference Sheet, *Tool and Manufacturing Engineer*, 47, December 1961:91.
FUKUI, S., KUDO, H., and SEINO, J., Study on Impact Extrusion Method for Nonferrous Metals, *Report Inst. Sci. and Tech., Univ. Tokyo*, 11:91 (1957).
GABLER, P., Cold Pressing and Extrusion of Aluminum in Light Engineering, *Aluminium*, 25:194 (1943).
GALLOWAY, D. F., Some Experiments on the Impact Extrusion of Aluminum, Copper and Steel, *Machinery* (London), 93, August 13, 1958:356.
—, Cold Extrusion—A Review of Some Technical Information and Current Practice, *Microtechnic*, 15, June 1961:85.
GARDNER, R. W., *Technique and Material for Cold Extrusion*, Society of Automotive Engineers, Preprint 36A, March-April 1958.
—, Evaluating the Producibility of Cold-Extruded Parts, *Tool Engineer*, 41, October 1958:105.
GOKYÛ, I., SUZUKI, H., *et al.* Studies on Impact Extrusion; ten reports in *Nippon Kinzoku Gakkai-Si*, vols. 18-20 (1954-56). Topics as follows:
1. Mainly on extrusion forces of zinc shells, 18:232

2. Mainly on defects and properties of extruded zinc shells, 18:236
3. Behavior of extruded zinc shells, 18:321
4. Mainly on extrusion forces and properties of aluminum shells, 18:375
5. On earing behavior of extruded aluminum shells, 18:379
6. On cold extrusion of square shells, 18:437
7. On cold impact extrusion forces, 19:351
8. On cold impact extrusion of brass, 19:354
9. On cold impact extrusion of steel, 19:575
10. On cold extrusion of magnesium and its alloys, 20:378.

GRIFFITHS, F., Cold Forming of Steels, *Machinery* (London), 97, December 7, 1960:1315.

HANES, R., Impact Extrusion; Some Preliminary Results of a Recent Research, *Sheet Metal Industries*, 26:89 (1949).

HATEBUR, B., Cold Upsetting and Cold Extrusion of Components on Transfer Cold Headers, *Wire Industry*, 28:775, 903 (1961).

HAUTTMANN, H. and PESSL, H. J., The Aluminum Bearing, Non-Aging Steels and Their Cold Extrusion, *Steel Processing*, 36, November 1950:557.

Heintz Manufacturing Company, *Report of Metallurgical Research on Cold Extrusion*: To Research and Development Division, Office of Chief of Ordnance, U.S. Department of the Army, September 1948; PB 135 708 (OTS) U.S. Department of Commerce.

HERRMANN, E., The Cold Extrusion of Aluminum, *Tz für praktische Metallbearbeitung*, 47, July 1937:528.

HOFFMAN, W. E. and TUTTLE, G. M., Cold Forming, *Tooling and Production*, 23, October 1957:91.

HOLLIS, W. S., Advanced Methods for Forming Metals, *Engineering*, 190, July 1, 1960:25.

HOWARD, F., Designing with Impact Extruded Components, *Engineering Materials and Design*, 1:26 (1958).

HOWARD, F., DENNISON, H. A. J., and ANGUS, N., Some Investigations into the Cold Extrusion of Steel, *Sheet Metal Industries*, 38, June 1961:403.

Impact Extrusion in Practice, *Light Metals*, 5:177 (1942).

JAMES D., Phosphate Coating and Lubricating Steel for Cold Extrusion, *Sheet Metal Industries*, 38, March 1961:171.

JOHNSON, E. and BISHOP, E., Die Steels for Cold Extrusion, *Ibid.*, 30, June 1953:490.

JOHNSON, W., Research into Some Metal Forming and Shaping Operations, *Ibid.*, 34, February 1957:121.

JONCK, R., The Influence of Composition and Structure on the Resistance of Steels to Deformation in the Forming of Solid Parts, *Revue de Metallurgie*, 59, January 1962:41.

KISTER, W., Investigation Concerning Rapid Aluminum Deformation *Schweizer Archiv*, 12:169 (1946).

Kudo, H., An Upper Bound Approach to Plane Strain Forging and Extrusion, *International Jour. Mech. Sci.*, 1, January 1960:57.

Kudo, H., Some Analytical and Experimental Studies of Axi-symmetric Cold Forging and Extrusion, International Jour. Mech. Sci., 2, October 1960:102.

Kuhne, H. J., *Der Kraftbedarf beim Fliespressen Zylindrischer Hohlteile*: Wilhelm Knapp, Halle, Germany, 1951.

Kunogi, M., New Method of Cold Extrusion of Steel, *Japan Soc. Mech. Engineers, Trans.*, 23, October 1957:742.

—, Cold Extrusion of Alloy Steels, *Ibid.*, 23, October 1957: 749.

Landberg, P., Remarks on Impact Extrusion, *Microtechnic*, 12, December 1958:301.

—, Plastic, Flow in Various Deformation Processes, *Metallbewerking*, 26, March 16, 1961:374.

Lang, R. F., Ansell, W. R., and Wilson, D. J., Impact Extrusion: German Practice, 1946, *B.I.O.S. Final Report No.* 1405, 1946.

Laue, K. and Hornauer, H., Limits of Deformation in Extrusion Pressing, *Zeit. Metallkunde*. 47:117 (1956).

Leland, J. F. and Helms, J. W., Troubles with Cold Extrusion May Be Traced to Lubrication, *Soc. Automotive Engineers, Jour.*, 62, November 1954:43.

Lloyd, T. E. and Kopecki, E. S., Cold Extrusion of Steel, *Iron Age*, 164, May 4, 1949:90.

Mabb, P., Impact Extrusion, *Machinery* (London), 76, February 9, 1950:192.

Mass, F., Cold Extrusion Molding, *Metall*, 7, September 1953: 689.

Matsuura, Y., *On Impact Extrusion*, Castings Research Laboratory Report No. 4, 1953, 48 (Waseda University, Tokyo).

Morgan, R. A. P., The Cold Extrusion of Steel, *Jour. Iron & Steel Inst.*, 193, November 1959:285.

—, The Cold Extrusion of Steel, *Metal Treatment*, 28:393, 451, 487 (1961).

Okell, R. E., Experiences in Commercial Production and Application of Cold Extruded Components, *Machinery* (London), 97, December 28, 1960:1457.

—, Metallurgical Aspects of the Cold Extrusion of Steel, *Metal Treatment*, 28, August 1961:309.

Patton, T. L., Impact Extrusion of Magnesium, Paper delivered at 7th Annual Meeting of the Magnesium Association, 1951.

Perry, R. W., Lubrication for Cold Forming, *Metals Engineering Quarterly*, February 1962, 50.

Phillips, R. M., High Speed Photos Reveal Extrusion Rate Phenomenon, *Iron Age*, 162, December 2, 1948:102.

Pineau, M., Backward Impact Extrusion, *La Pratique des Industries Mecaniques*, 37, June 1954:163.

Poulsen, S. C., Cold Extrusion of Steel, *Machinery* (London), 97, October 19, 1960:876.

Precision Techniques for Cold Extrusion, *Revue de Metallurgie* 58, September 1961:788.

PUGH, H. L. D., Cold Deformation of Metals: Part 1, Cold Forging, *Metal Treatment*, 27:189 (1960).

—, Cold Deformation of Metals: Part 2, Cold Extrusion, Metal Treatment, 27:231 (1960).

PUGH, H. L. D., WATKINS, M. T. and MCKENZIE, J., An Experimental Investigation into the Cold Extrusion of Steel, *Sheet Metal Industries*, 38, April 1961:253.

RIDINGS, J. B. and ANGUS, N. S., Experiments on the Reverse Impact Extrusion of Certain Metals, *Jour. Inst. Metals*, 90, December 1961:107.

ROLLINGER, W., The Application of the Cold Extrusion Process, *Aluminium*, 22:511 (1940).

RUSSELL, J. V., Steels for Cold Forming, *Metals Engineering Quarterly*, February 1962:49.

SABROFF, A. M., SANNICANDRO, R. A., and FROST, P. D., Design Considerations for Cold Extrusion of Titanium, Technical Paper 119, American Society of Tool Engineers, 1958.

SABROFF, A. M., HUBER, O. J. and FROST, P. D., Cold Extrusion of Unalloyed Titanium, *Trans. Amer. Soc. Mech. Engineers*, 80:124 (1958).

SABROFF, A. M., and FROST, P. D., A Comparison of Lubricants and Coatings for Cold Extruding Titanium, *ASLE Trans.*, 3, April 1960:61.

SACHS, G., A Review of Impact Extrusion and Some Related Processes, *Sheet Metal Industries*, 28, June 1951:533.

SHOEMAKER, J. D., Impact Extrusion of High Strength Aluminum, *Amer. Machinist*, July 6, 1953:129.

SIEBER, K., Developments on the Cold Flow Pressing and Extrusion of Steel, *Machinery* (London), 80, May 22, 1952:891.

STELLJES, H. A. J., On the Question of the Raw Materials and Their Properties in the Impact Extrusion of Light Metals (Al Alloys), *Metallforschung*, 2:341 (1947).

STOCK, A. J., A Review of Metalworking Lubricants for Stainless Steel, *Lubrication Engineering*, 17, December 1961:580.

TANNER, R. J. and JOHNSON, W., Temperature Distribution in Some Fast Metal Working Operations, *International Jour. Mech. Sci.*, 1, January 1960:28.

TILSLEY, R. and HOWARD, F., Cold Extrusion of Ferrous and Nonferrous Metals, *Machinery* (London), 97, December 7, 1960:1286.

—, Cold Extrusion of Ferrous and Nonferrous Materials, *Production Engineer*, 40, March 1961:176.

U.S. Department of Commerce (Office of Technical Services), *Cold Shaping of Steel:* Summary Report, PB 96704, 1947.

VAN ROUYEN, G. T. and BACKOFEN, W. A., A Study of Interface Friction in Plastic Compression, *International Jour. Mech. Sci.*, 1, January 1960:1.

VERSON, J., How to Design Practical Tooling for Cold Extrusion, *Amer. Machinist*, 101, October 7, 1957:129.
—, Practical Tooling for Cold Extrusion, *Metalworking Production*, 102, January 17, 1958:105.
VERSON, M. D., Tooling for Cold Steel Extrusion, Metalworking Production, 105, February 1, 1961:49.
—, Pressure Economy, *Sheet Metal Industries*, 38, July 1961:493.
VICTOR, M., Industrial Possibilities of Impact Extrusion, *Revue de L'Aluminium*, May 1957:535.
VON ZEERLEDER, A. and VON BERG, K., Relations of the Power Demands in Extrusion of Capsules of Pure Aluminum and Various Aluminum Alloys, *Schweizer Archiv*, 3:109 (1937).
WALLACE, J. F., Adiabatic Deformation in Impact Extrusion, *Jour. Inst. Metals*, 90:38 (1961-62).
WILLIAMS, C. G., Mechanism of the Action of Extreme Pressure Lubricants, *Proc. Royal Soc.* 212A, 1952:512.
WISTREICH J. G., Cold Working of Metals, *Jour. Iron & Steel Inst.*, May 1955:51.

Index

www.ingramcontent.com/pod-product-compliance
Ingram Content Group UK Ltd.
Pitfield, Milton Keynes, MK11 3LW, UK
UKHW041637031026
15587UKWH00001B/345

9 780820 600130